Environmental History

NEW PERSPECTIVES ON THE PAST

General Editor
R. I. Moore

Advisory Editors
Gerald Aylmer
Tanya Luhrmann
David Turley
Patrick Wormald

PUBLISHED

David Arnold Famine
James Casey The History of the Family
Patricia Crone Pre-Industrial Societies
James Fentress and Chris Wickham Social Memory
Ernest Gellner Nations and Nationalism
Ernest Gellner Reason and Culture
David Grigg The Transformation of Agriculture in the West
Richard Hodges Primitive and Peasant Markets*
Eugene Kamenka Bureaucracy
Edward Peters Torture
Jonathan Powis Aristocracy*
I. G. Simmons Environmental History

IN PREPARATION

David Arnold Environment and Culture
Richard Bonney Absolutism
David Gress The Balance of Power
John Mackenzie Hunting
David Turley Slavery

* now out of print

Environmental History

A Concise Introduction

I. G. Simmons

First published 1993

Blackwell Publishers
108 Cowley Road
Oxford OX4 1JF
UK

238 Main Street
Cambridge, Massachusetts 02142
USA

British Library Cataloguing in Publication Data
A CIP catalogue record for this book is available from the British Library.

Library of Congress Cataloging-in-Publication Data
Simmons, I. G. (Ian Gordon), 1937–
Environmental history: a concise introduction / I. G. Simmons.
p. cm. – (New perspectives on the past)
Includes bibliographical references and index.
ISBN 1-55786-445-4. – ISBN 1-55786-446-2 (pbk.)
1. Man – Influence on nature – History. 2. Man – Influence of environment – History. 3. Human ecology – History. I. Title. II. Series: New perspectives on the past (Basil Blackwell Publisher)
GF75.S58 1993 93-3257
304.2'09–dc20 CIP

Typeset in 11 on 13 pt Plantin
by Best-set Typesetter Ltd., Hong Kong
Printed in Great Britain by Page Bros. Ltd, Norwich

This book is printed on acid-free paper

Contents

Editor's Preface

Ignorance has many forms, and all of them are dangerous. In the nineteenth and twentieth centuries our chief effort has been to free ourselves from tradition and superstition in large questions, and from the error in small ones upon which they rest, by redefining the fields of knowledge and evolving in each the distinctive method appropriate for its cultivation. The achievement has been incalculable, but not without cost. As each new subject has developed a specialist vocabulary to permit rapid and precise reference to its own common and rapidly growing stock of ideas and discoveries, and in doing so come to require an ever greater depth of expertise from its specialists, scholars have been cut off by their own erudition not only from humanity at large but from the findings of workers in other fields, and even in other parts of their own. Isolation diminishes not only the usefulness but the soundness of their labours when energies are exclusively devoted to eliminating the small blemishes so embarrassingly obvious to the fellow professional on the next patch, instead of avoiding others which may loom much larger from, as it were, a more distant vantage point. Marc Bloch observed a contradiction in the attitudes of many historians: 'when it is a question of ascertaining whether or not some human act has really taken place, they cannot be too painstaking. If they proceed to the reasons for that act they are content with the merest appearance, ordinarily founded in one of those maxims of common-place psychology which are neither more nor less true than their opposites.' When historians peep across the fence they see their neighbours in psychology, as it might be, or literature, or sociology, just as complacent in relying on historical platitudes which are naive, simplistic or obsolete.

New Perspectives on the Past represents not a reaction against specialization, which would be a romantic absurdity, but an attempt to come to terms with it. The authors are specialists, of course, and their thought and conclusions rest on the foundation of distinguished professional research in different periods and fields. Here they will free themselves, as far as it is possible, from the restraints of field, region and period within which they ordinarily and necessarily work, to discuss problems simply as problems, without predefining them as part of 'history', 'politics' or 'economics'. They will write for specialists, because we are all specialists now, and for laymen, because we are all laymen.

There could hardly be a better example of a new perspective on the past than the one which is presented by the history of the environment. Its relevance to pressing contemporary anxieties is obvious, but it is also stimulating fundamental reassessments in many periods even of the remotest past, and in many parts of the world. It is truly interdisciplinary, with dimensions not only in history, archaeology and geography, but in the earth, biological and medical sciences, and born of one of the oldest inter-disciplinary alliances of modern academic times, that between history and geography, which themselves live astride the boundaries between the social sciences and, respectively, the humanities and the natural sciences. As the writings of Marc Bloch, Lucien Febvre and Fernand Braudel constantly remind us, the close relationship between history and geography in the French educational system was a principal inspiration of the comprehensive approach to understanding the past which they pioneered through the school of the *Annales*, and whose influence on the modern study of history has been incalculable. But the influence has been mutual, and geographers like historians have abandoned the simple-minded determinism that once seemed so seductive. It is very much in this tradition that Ian Simmons, as a geographer, argues so vigorously here that the critical issues with which environmental history confronts us are, in the end, issues of human culture and perception.

The pedigree of environmental history in the main lines of several traditional disciplines is important now precisely because the subject has acquired such urgent contemporary relevance. Like all complex relationships, that between humanity and the planet which it inhabits must be understood both as a system functioning now, and, from its beginning, as the product of development over time; both as what is observed and as what the observer constructs; both with passionate sympathy and with cool and sceptical detachment. It is a tall order, but these pages show how it may be fulfilled.

R. I. Moore

Illustrations

Tables

Preface

To write for a category of scholars and students outside one's own accustomed purlieus is a task which caused me some apprehension. The main problem, I decided quite early, was not knowing what they would or would not know. With geography students, experience provides a guide as to what events, places and processes they are likely to have heard of; with those of history, such guidance is lacking. If this book does not as it were go straight past its target group, then much credit is due to the Series Editor, R. I. Moore, for his advice on what conceptual and verbal items would be incomprehensible to potential readers. His comments on drafts at various stages have, both in general and in detail, been a great help.

In formulating the book, I have had in mind that students of history no less than other people will have received a daily diet of material about 'the environment' for some time. This book is not an attempt to persuade them to be Green activists nor indeed to be anti-nature technocrats. It is rather a suggestion that the human transformation of the globe over the last 10,000 years is worthy of the attention of historians at all levels. The undergraduate student can bring to the attention of his peers the fact that human alteration of ecological systems did not begin with the deployment of the bulldozer; the advanced scholar can entertain the notion that history is branded with the mark of radical uncertainty and that the complexity of natural systems has played a part in the outworking of such contingencies. One Talmudic text suggests that there were twenty-six attempts to create the World before the present Genesis and that this Creation was accompanied by the less-than-omniscient sounding words of God, 'Let's hope it works.'

So let's hope this book works, too. I have tried to take the reader from the immediately observable to the more abstract in a set of five chapters. We start with the commonplace observation that the development of human cultures has conferred the ability to change nature in varying degrees through time: a simple periodization is needed and some generalizations can be made about the human-induced alterations at each stage. I then describe what happens when humans change ecological systems: what do the natural sciences (and ecology in particular) tell us? Can they give a vocabulary, for example, that historians might find useful in thinking about the interactive context of human societies and their surroundings? In chapter 3 I deploy some of these concepts onto some of the major areas of the world land-cover map and delve into the history of the swathes of green, blue and brown and ask, 'were these always so; have humans changed them *in toto* (by removing forests in favour of cropland) or in detail, by replanting exploited woodland with an imported species?' Chapter 4 introduces the idea that nature has for long been subject to influences outside the scientific realm, and so the environmental history of political units (in particular those of England and Wales and of Japan) are made a focus for more detailed historical narrative. Inevitably, national attitudes to the natural world (or, more importantly, what is perceived as the natural world) arise, and so the book ends with a move towards the abstract and considers some of the complexities of human culture which surround the idea of nature and our behaviour towards it.

I also hope this book will both encourage the further development of the field of environmental history as practised in the humanities in North America and Europe, and stimulate multidisciplinary approaches. The field has already produced some distinguished scholars and I would be happy if more started out by being introduced to this book. The current debate over global change will be impoverished if it is not extended to a wide range of scholars in the humanities and social sciences, and historians have, I am certain, a great deal to contribute, not least in the avoidance of hubris.

The background to my own interest in this field is threefold: I have for thirty years been interested in one particular set of changes: those of the last hunter-gatherers on the uplands of England, and their story indeed surfaces from time to time in this volume. Beyond that, I tried to tell much of the story of chapter 1 at greater length in my 1989 volume for Blackwell: *Changing the Face of the Earth.* Additionally, the themes of the last chapter are further explored in my book, to appear in 1993, *Interpreting Nature: Cultural Constructions of the Environment* (Routledge, London). I am grateful for the invitation to bring them all together in this form.

Intellectually, I think it all started with the lectures on 'The Changing English Landscape' delivered at University College London when I was an undergraduate by Clifford Darby. Since, as Professor Sir Clifford Darby, OBE, FBA, he died this year (1992), it seems appropriate to recall them gratefully, both as stimulating material and as consummate lectures in themselves. After that, numerous people have been influential, but I should like to mention my longtime colleague at Durham, Brian Roberts, with whom discussion of historical processes in the landscape has always been interesting, often accompanied by good laughs and good pubs. As always, my other colleagues at Durham have been tolerant about those things I have not done and supportive about those which I have; a better set of colleagues with whom to confront these times of rapid change cannot be imagined. I have received much-valued (and needed) help with the linguistic section of chapter 5 from Dr Ralph Austin and Dr Raghu Char of this University (Arabic and Sanskrit respectively); Professor Pu Hanxin of Academica Sinica (Chinese); and Professor Satoh Masohiro from Osaka City University (Japanese). Many thanks to them for their time and interest.

In the preface to one of his books, David Harvey makes an acknowledgement to (among other musicians) Dimitri Shostakovich, and no doubt conclusions were drawn from that. I will do the same for Ralph Vaughan Williams and follow that up with the gist of his reported remark about the Fourth Symphony: 'I don't know if I like it, but it's what I meant.'

I. G. Simmons
Durham

Acknowledgements

The quotations in the chapter titles are all from poems by W.H. Auden, and are reproduced from *Collected Shorter Poems* (Faber and Faber, 1966) by kind permission.

Chapter 1: 'Not the sweet home that it looks', from *In Praise of Limestone*.

Chapter 2: 'Might-be Maps', from *Makers of History*.

Chapter 3: 'A culture is no better than its woods', from *Woods*.

Chapter 4: 'Not dream of islands', from *Paysage moralisé*.

Chapter 5: 'Our good landscapes be but lies', from *Ode to Gaea*.

1

'*Not the sweet home that it looks*': A History of the World in Only Five Chapters

To provide a context for other kinds of human history, and to emplace more firmly the idea that nature and humanity have engaged in mutual interactions (rather than nature simply providing, passively, a stage for human accomplishments), we shall start by attempting a broad overview of the last 10,000 years in terms of the environmental relations of human societies. This will necessarily deal with the impacts caused by people and their technologies as well as the constraints posed by environmental factors. To do this, some kind of periodization is needed, which always carries its own risks. In this case, the risk is that the characteristics of any stage always carry over somewhere or other into the next; that is, there are regions which lag behind others, just as there are those which are ahead of their time in being precursors of a more general subsequent change. But with such caveats in mind, we can adapt the notion of stages of a *cultural ecology* as set out by W. I. Thompson[1] and somewhat elaborated here.

This scheme identifies a number of periods in which human cultures and environmental interactions in various parts of the world had a sufficient unity to enable distinctions to be made and demarcated. They comprise:

[1] W. I. Thompson, *Imaginary Landscapes. Making worlds of myth and science* (St Martin's Press, New York, 1989).

1 *Hunting-gathering and early agriculture.* All humans employed this first type of economy until they learned to domesticate plants and animals. This pattern was first fully established about 7500 BC in south-western Asia, though hunter-gatherers have persisted in gradually diminishing numbers until virtually the present day. In general, the hunters manipulated the environment less than later cultures and adapted closely to its given features.

2 *Riverine civilizations.* These were the great irrigation-based economies, for example of the Nile and Mesopotamia, surrounded from about 4000 BC by nomadic pastoralism in the dryer and more mountainous habitats, and lasting until about the first century AD. Through technology these civilizations attempted to free themselves from the some of the constraints of a season virtually without rainfall.

3 *Agricultural empires.* From 500 BC until the eve of the full industrial revolution (*c.* AD 1800) there were a number of city-dominated core areas of the world, in some cases political empires as well as commercial ones. Each might be said to have had a core area, a middle zone of considerable influence, and a peripheral area where any changes (of any kind) were sporadic in time and space. Many adopted technology to overcome environmental barriers to greater production: for example, water storage, terracing and selective breeding.

4 *The Atlantic-industrial era.* From about AD 1800 to the present day, a belt of cities from Chicago to Beirut, plus a few outlining the shores of Asia as far as Tokyo, have provided the core area of an economic mode based largely on energy derived from fossil fuels. This has been the period of the greatest impact of our species upon its surroundings. It has also provided many examples of societies isolating themselves from their natural environment, for instance, through air conditioning and long industrially-based food chains.

5 *The Pacific-global era.* Since the 1960s there has been a shift in emphasis to the Pacific Basin as a primary focus of the industrial economy, but at the same time a true globalization of communications has facilitated the emergence of an inter-

dependent world economy, for example in the sphere of operations of finance and of multinational corporations. This has been accompanied by a shift in consciousness towards a global outlook, one sign of which is a resurgence of interest in diversities of lifestyle based on local environmental characteristics: this is known as 'bioregionalism'.

This book is mainly concerned with the effects of humans upon their environments, rather than with a discussion of the merits of environmental determinism. Let us now go on to consider each of these phases and the types and distribution of ecological change wrought by humans within it.

Hunter-gatherers and early agriculture

At 10,000 BC, the populations of *Homo sapiens* were all hunter-gatherers. As such, they were a very successful group, for few other animal species could be found in as wide a diversity of habitats. Apart from areas of year-long snow and ice, and the highest slopes of mountains, the only uninhabited places were some remote islands (Hawaii was only colonized by humans in about AD 400), the deepest recesses of deserts and perhaps some steppes and grasslands. Otherwise hunter-gatherers were able to occupy the full range of habitats available in the wake of the retreating ice and its global effects, and also to adapt to rapidly changing ecologies as much of the world underwent climatic amelioration. Hence, before being displaced by agriculture, conquest or disease, hunter-gatherers were found from very high latitudes in the northern hemisphere, through the lowland tropics, to Tierra del Fuego, that is from about 75°N to 65°S. This great ecumene was achieved without sophisticated technology, for the essence of survival was usually seasonal movement, and so an abundance of possessions (or of children unable to walk) was an encumbrance to be avoided. The technology centred around the enhancement of the body's energy by concentrating it down the narrow trajectory of a spear, arrow or blowpipe dart, and the harnessing of stored solar energy in the form of a domesticated animal aid to hunting (the dog), and as controlled fire. With diets that usually contained as much plant material as was necessary and as much animal flesh as could be obtained, hunter-gatherers largely peopled the globe.

For our theme, the question to be addressed is, 'given the low population densities and seasonal movements, did hunter-gatherers affect the ecology of their environments, and if so, were any changes only temporary or might they become permanent: did those people create humanized landscapes?' In the latter endeavour, if it happened, then the possession of controllable fire would probably be very important, as would the flammability of the habitat. But there might be other ways of creating such changes, for instance by reducing or eliminating a key animal species, with consequent direct or indirect effects upon the rest of the relevant ecosystem. W. Schüle[2] has even advanced the hypothesis that such effects may have included global climate.

In some places, there seem to have been irreversible effects. The earliest of these is termed the 'Pleistocene overkill' hypothesis and refers to the rates of extinction of large mammal herbivores during the last phases of the Pleistocene and the early Holocene, when some localities were receiving their first human populations. Between 12,000 and 10,000 BP some 200 genera of mammal herbivores with an adult weight of ⩾50 kg (the so-called 'megafauna') became extinct; the phenemenon is most closely marked in North America, where two-thirds of the large mammal fauna disappear from the fossil record. Here, the animals involved include three genera of elephants, giant armadillos, pangolins and anteaters, fifteen genera of ungulates (deer and antelopes) and various large rodents and carnivores. If the conventional time (starting about 12,000 BP) for the introduction of humans into the New World via the Bering Strait is accepted, then these extinctions look as if climatically stressed animal populations were made extinct in the course of a southward expansion of human hunters. Climatic change by itself is for some palaeontologists a preferred explanation, but the human linkages are given some additional credibility by analogous events in other places. In New Zealand, the moa bird (a flightless emu-like creature) succumbed within a few hundred years of the first peopling of the islands; in

[2] W. Schüle, 'Anthropogenic trigger effects on Pleistocene climate?', *Global Ecology and Biogeography Letters*, 2 (1992), pp. 33–6.

Madagascar the same happened to a similar bird called *Aepyornis* and to a pygmy hippopotamus, and in Java to populations of pygmy elephants. By contrast, the picture in Eurasia and Africa is one of much lower rates of extinction: the perishing of 73 per cent of the genera of the terrestrial megafauna in North America (80 per cent in South America and 86 per cent in Australia) was scarcely matched by 14 per cent in Africa, and Eurasia lost only nine species, all except one of which were animals of the cold steppe but which had survived in large numbers until about 12,000–10,800 BP, when vegetation change was rapid. Humans in Africa and Eurasia may well have been less of an influence than elsewhere.

Later examples of human impact of a permanent nature upon habitats are well documented. In Australasia, for instance, the aboriginal inhabitants used fire for hunting, land clearance and communications (they said that fire was used to 'clean up' a landscape), with an estimated 5000 bush fires per year in Australia in the early nineteenth century. Fire in Tasmania was used to convert the mature southern beech forest to drought-tolerant forest and thence to scrub, fern heath and tussock grassland. The environmental mosaic which resulted encouraged wallabies, bandicoots and possums, all of which were edible. In New Zealand the bracken fern was a starchy staple food of the Maori, and fire was regularly used in the period AD 1000–1400 to encourage its growth and spread; as a result, evergreen forest became grassland and scrub. Along with the mature forests forty species of bird disappeared, including twenty species of moas. On the coasts both elephant and fur seals disappeared except in remote areas.

The last gatherer cultures of upland England are discussed in more detail in chapter 2. Here we may note that it seems that they either enlarged forest clearings or made them *de novo* so that the resulting scrub would attract more deer than the mature high forest of oak and other deciduous trees. In this, fire was an important tool. Left to themselves, such clearings might well return to forest once again, though probably not with exactly the same species composition. But held open for some decades, the clearings experienced a different water regime from the forests. The woodland canopy in summer intercepts a good deal of rainfall and allows it to evaporate straight back to the atmosphere; at this

Figure 1.1 A relict hunter-gatherer landscape in upland Britain. The tree remains date from a time when forest grew back over shallow peat after the last hunter-gatherers ceased burning the vegetation. The basic openness of the landscape has its origins in these pre-agricultural times.
(*Photograph*: I. G. Simmons.)

season too, the roots of the trees act as water-pumps in removing water from the soil and making it available for transpiration from the trees' leaves. If there are no trees, then much more water remains in the soils; eventually waterlogged and acidic soils begin to accumulate only partially decayed organic litter. Peat then begins to build up into a blanket which is in some places still there, although much eroded where grazing and acid precipitation have destroyed the carpet of living plants.

But equally, there are accounts of hunter-gatherers who appear to have had no impact upon the environment. The Inuit of Arctic North America, for example, seem to have had no effect on the populations of caribou or of sea mammals, at any rate not until the advent of the rifle, snowmobile and outboard motor. There is no evidence that occupants of the African savannahs affected the population dynamics of the great herds of herbivores, some of which still exist in spite of encroachment upon their habitat and poaching. Traditional reindeer hunting in northern Europe lasted many hundreds of years in spite of being non-selective and indeed wasteful. The Great Basin Shoshone needed to allow antelope populations eight years to recover from one cooperative hunt. In such circumstances, human societies would not have survived at all if they had imperilled all their resources at once and so there must have been some continuing source of subsistence.

Scholars are thus divided as to whether hunter-gatherers were archetypally conservationist (and hence perhaps good 'Green' role-models) or simply too thin on the ground to produce the sort of environmental impact exhibited by later economies. Some argue that the groups had no conception of Man as distinct from individuals, and thus a generic responsibility was very unlikely. Examples of groups such as the Hadza of Tanzania, who killed whenever possible, reinforce this view. Others, by contrast, point to definite conservation practices and cultural sanctions. The G/wi bushmen of Namibia angered their Supreme Being if they killed or gathered any more than was needed; the Montagnais of the northern forests randomized their kill areas through their divination practices, and the Aranda of Australia declared all sites within two kilometres of their settlement sacred, and also had totemic prohibitions that may have provided differential refuge for particular species; the Birhor hunters of northern India allowed areas to lie 'fallow' for one to four years.

This division of opinion is quite sharply contested in North America, where the social and political rehabilitation of the remaining aboriginal inhabitants is being attempted and where a few well-known pronouncements (such as that of Chief Luther Standing Bear[3]) set a tone which helps that cause. It is possible

[3] Reproduced in, for example, T. C. McLuhan (ed.), *Touch the Earth* (Abacus, London, 1973). There has been some recent suggestion that these and other ecologically tender statements have been 'enhanced' in their translation to the modern environmentalist literature.

that both environmental tenderness and unsustainable culling were characteristic of the same group: the one at times of stability and the other at times of stress. The picture is one of variability and contingency rather than one driven by iron laws. However, any picture of hunter-gatherers simply as responsive children of nature living solely off a provident usufruct is part of a myth of a Golden Age.

Why agriculture became a successful alternative is unknown, though hypotheses are plentiful. It looks as if full agriculture, with dependence upon domesticated animals and plants, was preceded by a period of husbandry when wild biota (i.e. all living organisms) were quite intensively exploited but without any intention to breed for particular characteristics. Late Pleistocene climatic change in south-western Asia, for example, may have resulted in areas of open woodland with quite dense stands of wild grasses in the gaps. These stands could well have been harvested for their seeds and a human group might return to them regularly each year and even protect them near harvest-time from bird or mammal predation. Where feasible, the diversion of a watercourse to improve the swelling of the seed might have been part of the husbandry repertoire, along the lines of the flood which rose from the ground and watered the earth in the Garden of Eden.

With animals, close contact is possible through at least two routes. One is the taking of young and bringing them up within the human group, initially perhaps as pets for children; another is the way in which a species like the pig would be attracted to settlements as a scavenger. Yet a third envisages cattle being corralled for use in religious ceremonies on account of the lunate curve of their horns. This below-the-horizon husbandry might have lasted for thousands of years or it could have tipped within a few generations into conscious selection of plant seeds and animals' parents which brings about that fundamental element of domestication which resides in controlling the genetics of an animal or plant population. Somewhere on the flanks of the mountains of present-day Iran, Turkey, Syria, Iraq and Palestine there occurred the combination of cereal agriculture (in which varieties of wheat were initially the most important carbohydrate source) and animal husbandry based on domesticated sheep and cattle, with the latter species confined to areas able to provide water nearly every day. To manage these systems, sedentism was essential, even though some men might go out to hunt wild animals, and women to gather wild plant material.

This region was the source area for cultivation-based economies throughout Europe, western Asia and the Mediterranean, but not the only one: it seems as if domestication was independently arrived at in several other places. During the period 7000–4000 BC maize and the potato appeared in Mesoamerica and on the slopes of the Andes, rice and chickens in south-east Asia and sorghum and yams in Africa. Once intercontinental travel became successful, these source areas then cross-fertilized one another so that by the sixteenth century AD crop plants and weeds, animals and their parasites were being implanted in many new cultural and natural environments. In the period to about 500 BC, though, the environmental impact of this type of agriculture was essentially low, being confined mostly to obliteration round a settlement and the consequent mix of diversification and simplification attendant upon the small-scale crop agriculture and animal-keeping being practised. But the success of the novel economy is undoubted and the hunter-gatherer populations declined except in areas where agriculture and animal-raising could not be carried out – where it was too high, too dry or too cold. So in the Middle Ages of Europe, for instance, hunter-gatherers were confined to what are usually termed 'marginal environments': marginal to agriculture, that is. It was in these situations that most travellers and early ethnologists found the remaining populations which they described; it is as well to remember the dangers of extrapolating back to prehistory the cultural customs (including the environmental attitudes) of these remnant populations.

In summary, therefore, the world at say 5000 BC was certainly affected by those millennia of hunter-gatherers and early agriculturalists. But their impress was in the main light and often temporary and there must have been many places that escaped altogether, notably the seas. If viewed from the air, then the effect of human groups would certainly have been present; but the world was still largely a wild place.

Riverine agriculture

Those who developed agriculture in the hill fringes of south-west Asia must soon have realized that crops would not grow beyond certain limits which were related to the availablity of water. In our terms, the boundary of rain-fed agriculture coincided with the 300 mm isohyet. Thus it seems to us a natural progression for the techniques of irrigation to be developed in a series of villages on the dry side of the boundary. The spreading of water by simple ditching networks across fields made on alluvial fans was probably the beginning. Then, when they found that this was efficacious, cultivators were able to colonize the riverine plains of Mesopotamia and Egypt. Similar developments happened in the valleys of the Indus (probably influenced by events to the west) and around the Yellow River of China, probably in isolation. All these lasted in varying condition from about 5000 BC until around the middle of the first millennium BC: some became part of greater units, as in China and India; Mesopotamia relapsed into steppe; and Egypt stayed much the same until technology allowed the damming of the Nile on a large scale.

The importance of these riverine cultures in human history is often stressed; for our purposes it is the technological achievements which need to be highlighted, and in particular the ways in which water could be controlled and, especially, stored in large quantities in order to extend the growing seasons of cereals. These represent only the second major alterations of the physiognomy of the natural world which were made consciously and to last (the application of fire to semi-arid grassland and savannah, as mentioned earlier, being the first). Their material success provided energy surpluses that allowed society to become stratified and to divert effort previously required for subsistence into, for example, hunting for pleasure, garden construction, elaborate funerary practices and warfare.

The wildlands which were converted into these great granaries were seasonally inundated plains interspersed with permanent lakes and swamps. They were vegetated by papyrus, sedge and reeds, grasslands, savannah-like tracts with *Acacia* trees, and woodland dominated by trees able to tolerate seasonal waterlogging. Waterfowl and fish abounded. So, presumably, did the crocodile: perhaps it was a large male that Ezekiel (29:3) had in mind when he compared the Pharaoh to 'the great dragon that lieth in the midst of his rivers'.[4] In terms of more recent habitats, we might think of the Florida Everglades, or the Iraqi marshes as pictured during the Gulf War in 1991. Transformation required the twin processes of drainage and clearance: in Egypt, ditches not only drained water off the alluvial plain but also led it into storage ponds whence it was released onto the fields. Trees and bushes were felled, and the uneven ground surface was broken up by hoes and ploughed with ox teams, and eventually sown. Control of the water in the reclaimed land requires straight ditches and canals both to distribute water in the dry season and to drain it in the wet. For the former, though, lifting devices for water are essential. The success and stability of the systems led to bigger projects. One queen of Assyria had a storage lake built which was 75 kilometres in circumference: its diameter would have been about the length of the Dornoch Firth or the Thames between Henley and Windsor, about 25 kilometres. If the main river and its tributaries are likely to flood unpredictably (which was certainly true of the Euphrates and Tigris), then natural embankments must be strengthened and supplemented, and kept in good repair against the depradations of burrowing animals like water-rats. The advances of irrigation technology can be measured from the building of a masonry dam on the Uzaym River in the first millennium AD which was 170 metres long with a 15-metre crest.

These new agricultural systems seem to have been stable and successful. No doubt some food was obtained by trade, but

[4] But perhaps, less flatteringly, it was a hippo. The New English Bible reads 'monster' for dragon, and then also at 32:2,

> You were like a monster in the waters of the Nile
> scattering the waters with its snout
> Churning the water with its feet
> and fouling the streams

which sounds more like a hippopotamus, with that creature's defecatory habits. Either animal is a nuisance to cultivators: the crocodile for pretty obvious reasons, the hippo because the herds come out at night and eat grasses within a few hundred metres of the river. At the very least they would trample the crops.

staples and unpreservable items must have come from the local agroecosystems. Thus, areas of land in and near the towns were converted to gardens where the production of edible items (grapes, figs, dates, herbs and fish) was intensive. In fourth-millennium Mesopotamia there are depictions of harnessed cattle which suggest that there may have been an integrated crop-livestock pattern. The introduction of more fat into human diets may have brought down the age of menarche and thus increased rates of population growth, in turn creating the pressure for more land conversion. It has been calculated for Egypt in the eighteenth century AD that an energy input of 10 Gigajoules per hectare per year (GJ/ha^{-1}/yr^{-1}) by animals and men was followed by an output of 19 GJ/ha^{-1}/yr^{-1} of edible energy for humans together with 62 animal feed days of clover and 122 feed days of straw; a human population density of 4.5 per hectare was supported. If that were true of the Old Kingdom (2686–2160 BC), then a decent diet could have been delivered to the population by two hours per day of adult male labour, thus freeing a labour surplus for monumental projects.

The ecology appears to be simple: water is supplied for a time outside the natural rainfall period and so the growing season is extended. Nutrients are replaced in the silt which, in the Nile at least, covered the land every year and which elsewhere was probably spread on the land, and by animal manure. There seems to be no record in the Nile and Mesopotamia of the Chinese practice of composting human excreta for later spreading on the fields.[5] It is not unconditionally stable, for good water management is essential. In semi-arid climates, the excess of evaporation over precipitation draws water up through the soils by capillary action. Thus minerals are translocated upwards in the soil profile and can salinify the soils to the point where they will grow only crops which are tolerant of very high salt levels. So soils have to be flushed downwards regularly but not too zealously: in Egypt the Nile undertook this task yearly; elsewhere it had to be arranged by

[5] Though we might wonder why the Nile turned to blood during one of the plagues of Egypt (Exodus 7:20): this sounds like a red tide, which is an explosion of microscopic organisms associated with warm water and an excess of nitrogen and phosphorus.

whoever was in charge of the water-works. Likewise, irrigation and drainage systems are only as good as their ditch systems: if they are choked, then water can get neither to nor from the croplands.

There is evidence for Mesopotamia that during periods of political division and unrest these tasks were not attended to; the problems may also have been made worse by increasing silt burdens in the drainage system. These latter could have been brought about by intensification of cropping to feed extra mouths or might have resulted from erosion of the surrounding uplands; this in turn might have resulted from some forms of pastoralism, discussed below. By at least 1700 BC, severe salinity problems were growing at the lower ends of some of the canal systems of Mesopotamia, and the more the Tigris water was used, the more the silting became intractable. This coincides with an estimated population peak of 630,000 in this basin in 1900 BC, followed by a 300-year decline down to 270,000 in 1600 BC. Only under the aegis of the Persian empire did the numbers rise again significantly. Searchers (and especially those of Green persuasion) for single-factor explanations of the decline of particular societies may point to these trends as critical in the later history of the Tigris-Euphrates basin, though we could mention that it has in recent years supported a density of 139 persons per square kilometre as against a maximum of 11.4 per square kilometre in 1900 BC.

These land and water transformations are relatively sudden and complete, though they may not last for ever. The economy of pastoralism is rather a contrast in its transforming power. The origins of pastoralism are usually taken to be later than that of settled agriculture, beginning perhaps about 4500 BC and originally marginal to the riverine civilizations of south-west and south Asia. The virtues of pastoralism as an economy are simple: choice of the right animal and access to the right seasonal pastures can turn plants inedible to humans into nourishing products like meat, milk and blood as well as useful materials such as horn, hides and dung. Surplus animals can always be traded with valley-dwellers anxious for a change of diet, and if a long stop is possible at one place then some cultivation can be undertaken as well. Somehow, this combination of constant movement with the dangers of marginality have made the life of the pastoralist, from

Genghis Khan through the cowboy to T. E. Lawrence's friends, seem romantic to today's western eyes.

The ecology of pastoralism has some complications, arrayed firstly around water. By definition, the lands marginal to the great river basins under discussion are likely to be short of water, at least seasonally. This means, for example, that cattle (and indeed the horse) are less likely to provide the basis of such an economy since they have to drink free water every day. Hence the attraction of animals which can get much of their water from vegetation or can store water in fats. Sheep, goats and the camel predominated in the south-west Asian group, though elsewhere in the world the llama, yak and reindeer have occupied that niche. Secondly, a herd of beasts confined to a small area will quickly exhaust the forage. Constant movement to fresh woods (if available: most animals will browse as well as graze) and pastures new is therefore a necessity; and if water is not an absolute requirement for the domesticated mammals, then it is for the humans. In the semi-arid zones, though, the vegetation may not simply re-grow: shrubs may have had too much of their twig zone removed at flowering time; herbs may use up all their underground food reserves trying to re-grow and set seed. So the vegetation will change under anything but the lightest pastoralism. In general, plants inedible to domesticated herbivores (those with spines or toxic leaves, for example) will thrive, but few others will do so. Thus the net amount of vegetation per unit area (the *biomass*) will decline and there will be more bare soil. In sum, the system will become more *xeric* or desert-like, and bare soil will yield large quantities of silt to the run-off and threaten floods and silting downstream. So the happy eater of bought-in mutton chops in Uruk is, unbeknownst to him, undermining the system which brings him his rather more crucial bread.

The contribution of the riverine societies to later human development was very great and their echoes are still plentiful, not least in our 60-unit divisions of space and time. It is no surprise, then, that the alliance of such intellects with centralized socio-political control should have been able to carry out large-scale land transformations and to control water over wide areas; the contrast with their immediate forebears is strong. If some (though not most) of these societies suffered a degree of breakdown, then it is pertinent to compare their tenure of their environments with the current state and longevity of post-nineteenth-century industrial societies.

Empires based on agriculture

From about 500 BC until AD 1700, a major feature of the world's economy was the existence of city-centred regions with a degree of central commercial and political control. In some cases, this was sufficient for the label 'empire' to be applied, as with imperial Rome, or Spain at the end of the sixteenth century. But in every case there was a core focused on the major city or cities, a middle zone where the influence of the centre was strong and a periphery, perhaps under nominal control but which was largely untouched by the empire even if altered by its native inhabitants. About fifteen of these can be readily identified.

The transformations of the core zone could be many and varied, but naturally enough any account must first focus upon food production, in which agriculture of various types predominates. Although shifting cultivation was still common in this world, it could never have been a sustaining force in the type of empire we are discussing (an exception is sometimes made for the Classic Maya, though it seems to have been only part of their subsistence repertoire), and attention must concentrate on permanent agriculture – for it looks as if cereal production was an essential feature of the sustenance of any urban centre in an empire – and the ways in which it can be made sustainable and in which it can be intensified in the face of higher demand. As in previous eras, there are two basic types: rainfed and irrigated.

Rainfed agriculture can produce all kinds of crops, but a basic energy source must always be among them. Hence the importance of cereals in so many of the core areas and, in areas unsuited to their growth, starchy roots such as the potato, yams and manioc. To sustain yield year after year (or on a shorter cycle in the tropics) several conditions have to be met, some of which have implications for the surrounding environment. The first, as ever, is that there should be a net energy surplus to all the humans in the appropriate social group (and if there is to be trade or forced export an even greater positive balance); this means that the cultivators are constantly seeking ways of keeping their energy input down. One is by living close to the fields so that as little energy as possible is spent on getting to work; another is the employment of domesticated animals to plough, haul loads, thresh crops and draw water. These beasts must of course be fed, but many can live off material inedible to humans like straw and chaff, or be fed off the interstices of the landscape – from patches of uncultivable grass or scrub, for example, or in forests marginal to the cultivated area. But their chief role in most rainfed systems is as a source of nutrients for the soil. Continued cropping means that a soil has natural nutrient input only from subsoil weathering, a slow process in most regions, and rainfall, which is also very variable by region. So techniques of replenishment of nutrients are critical to continued output: manuring from domesticated animals, fallowing, green manures (such as forest litter or seaweed) and nitrogen-fixing crops are all well known. The most significant environmentally, however, are the animals, because they have to be fed and for part of the time at least this may be in surrounding ecosystems which they gradually transform, like any species kept in a restricted space. So cattle may be essential to keeping up cereal production, but only at the expense of the regeneration of woodland, for example. Where horses are kept as draught animals or as mounts for superior social classes, then good quality grassland must be provided: in medieval central Europe the saying ran that 'horses eat people'. Keeping up the fertility of the fields might mean the artificial maintenance of slopes by carrying back upslope any soil washed downwards and also scouring the countryside for almost any organic matter that could be added to soils. Soot and pigeon dung are further examples: we can only hope that the pigeons did not feed on the corn. Inorganic materials like chalky marl were also used: this practice was noted by a visitor to Roman Britain.

It was also essential to preserve the physical structure of the soil, and in that task the addition of organic matter such as animal dung-straw mixtures was important; so was the prevention of wholesale downward movement under the influences of gravity and water flow. Soil lost from fields has three main downslope effects: it accretes in the valleys, changing their cross-profile; it is transported by rivers, adding to their flood potential; and it adds to delta and lagoon formation off coasts, with a potential for enlarging the habitat of malarial mosquitoes. The terrace must have been invented to combat this. This simple construction is perhaps the world's most widespread visible symbol of the agriculturalists' imprint on the earth, for it was employed not only in a variety of regions outwards from the tropics to the temperate zones, but also for irrigated agriculture, of which the wet-rice field is the outstanding example.

Extra demands upon the rain-fed systems have important environmental consequences. The first of these is simply to extend the area of cropland at the expense of another set of ecological systems: the forest is often the nearest and most obvious land bank. Wetlands can often be drained as in Europe and China, and land reclaimed from salt marshes by embankment. In the Europe of the sixteenth and seventeenth centuries a more complex set of processes involved the reduction of the fallow and the substitution of three- and four-course rotation including turnips and clover; the extinction of common rights and the consolidation of strips; and the breeding of more specialized animals. One at least of these was laden with environmental change, for the new fields thus created were often bordered with hedgerows which provided habitats for plant and animal species for whom the great open fields were inimical habitats: we can imagine the partridge and the fox, for example, growing in numbers. As knowledge accumulated and was more readily transmitted via the printed book, plant and animal breeding became more exact and so varieties could be generated which performed best in given conditions: for example, wheat that was tolerant of ever dryer conditions allowed farming to be pushed further and further west in Canada. In the 100 years after 1750, crop productivity (i.e. yield/unit area) doubled in England and increased by 60 per cent in France. However, the same kinds of process applied in the nineteenth century in the Black Earth region of Russia produced reports of 2 per cent of the land suffering from gulley erosion, with some ravines cutting headward at a rate of 200 metres per year.

The water relations of rainfed agriculture are often subjected to periods of excess. In irrigation, the problem is to apply the right amount of water at the right time, a time which is more or less by definition a period of natural dryness and so storage is often a

Figure 1.2 A solar-powered agricultural landscape in south China. The crops are watered by gravity-fed irrigation and nourished by the recycling of many types of organic matter, including human excreta.
(*Photograph*: I. G. Simmons.)

necessary concomitant. Thus the necessity for constructions to store, deliver and drain. Such works are generally manipulations of the environment and may well include terracing. The epitome of this system is the growth of rice in conditions which require it to be kept under 100–150 millimetres of slowly moving water for three-quarters of its growing period: *padi*. Many valleys have been converted to this land use, but it is at its most spectacular when flights of terraces are grafted onto a steep hillside as for instance in insular south-east Asia.

The nutrient content of the *padi* is sometimes kept up by the water itself, for if it drains from a volcanic slope then it may carry quite high loads of dissolved minerals. If not (or in addition), the warm water is conducive to the growth of blue-green algae which fix nitrogen, and the stubble from the last crop also decays and releases its nutrient content. Even so, farmers often add ten to fifteen cartloads per hectare of manure, human nightsoil, alluvium, ashes, soot, straw and any green compost that is available. In south Asia, some twenty-six soil management techniques and twenty-three forms of water management have been listed for *padi* areas, but the control of slopes via bunds and terraces is by far the most important and is also responsible for the most noticeable realignment of entire blocks of landscape.

The extension of such a system requires a very heavy capital input of energy in the form of human labour, and also of the cultural capital that accompanies the water control: in India in AD 75–100 the extension of *padi* required a kind of ecocultural package that comprised the clearance of forest, the construction of the tank to store the water, the laying out of the fields and the establishment of a settlement together with a temple, whose priests were the water control authorities. Intensification of existing cropland may be an attractive option. If fast-growing varieties of rice can be found, then better water control and more nutrients may enable another crop to be taken during the year. Closer planting may be possible, and more labour (and here children are very handy) may scare away scavenging birds. Better storage structures may mean a lower loss to rats immediately after the harvest. On the other hand, more frequent drying-out of the *padi* between crops may interrupt the life cycles of the fish, frogs and crustaceans which had been a tasty addition to the diet.

East Asian *padi* was not the only form of irrigation agriculture to transform the earth. After the fall of the Roman empire the Muslims rehabilitated many defunct irrigation systems in the lands which they conquered, bringing both new technology and new crops. Of the period AD 900–1000, one scholar remarks that in the Arab lands 'there was hardly a river, stream, spring, known aquifer or predictable flood that was not fully exploited.'[6] In the Mediterranean lands, irrigation added a whole new growing season and to fill it there were such novel crops as cotton, sugar cane, aubergine, water melon and sorghum; in the tenth century

[6] A. M. Watson, *Agricultural Innovation in the Early Islamic World: The diffusion of crops and farming techniques 700–1100* (Cambridge University Press, Cambridge, 1983).

AD the orange came to Seville (as did the lime and the lemon, all originally from south China and south-east Asia); and above all, there came rice. Summer residents of Chiantishire eating their lunch-time *risotto* looking onto the west front of the local Duomo might add a certain historico-religious irony to their bill.

Once interchange between environments or conquest of territory took place, then the type of species exchange described above for Islam and the Mediterranean became possible on a large scale. The early ability of the Portuguese, for example, to navigate any ocean passage meant that in the 1400s they introduced maize, sugar cane, bananas and the grape to West Africa and the Atlantic islands; thereafter they were imitated by most of the other empires, so soon after 1492 the maize and potato plants were introduced to Europe; horses and cattle went the other way. China supplemented rice carbohydrate with groundnuts and sweet potatoes in the seventeenth and eighteenth centuries, for they would grow on hillsides above the rice limit. Examples of economic crops are legion, though less notice has been taken of the accompanying biota. In almost every case, the transfer of a plant crop meant that typical weeds and pests of the crop went as well, and in the case of animals both endo- and ecto-parasites. Weedy plants which are able to thrive in almost any broken ground in many climates became world travellers: the best-known case is that of the European broad-leaved plantain (*Plantago major*) which was dubbed 'the white man's footprint'. The increased movement of people and their ships also meant that bacterial diseases might break out from source areas and be transmitted over large areas. So when we think today of the unpredictable consequences of technological change, we are tapping into a long history: the caravel out of Lisboa was yesterday's Boeing 747.

Agriculture means the exertion of control over the genotypes of plants and animals: the process of domestication. A significant point along that road comes when a species' success in growth and reproduction is dependent upon human intervention and when it would not survive in the wild. Thus domesticated cereals need planting out at the right time and perhaps spacing; they need protection from birds and, if possible, from fungal and other diseases. Left to themselves, such species would probably become extinct quite quickly. Animals might very well manage better, but many strains would lose innate defence behaviours against predators and so be relatively easy prey. It is difficult to imagine some of today's more pampered dog breeds succeeding in a world without humans. Empire-based systems of agriculture also had a significant effect upon nutrient flows, for they dislocated the sites of production and consumption: whereas in subsistence agriculture much of a crop is consumed very near the place where it is produced, export over long distances means that some at least of the nutrients are removed from the local agro-ecosystem and must be replaced. Hence the search for other sources of minerals and the importance of animals in, for example, eating upslope vegetation to convert to manure for the valley crops; and hence, come the nineteenth century (to anticipate our story), the immediate popularity of cheap, easily transported, bagged chemical fertilizers.

The production of food is not the only economic process which can bring about environmental transformation. The forests of the world are another set of sites of use which reflect deflections of their ecologies at human hands. As mentioned above, they act as a land bank for agriculturalists because their nutrient cycles are normally tight and so when first cleared the fertility of the soils is high relative to land which has been cleared for some time, especially in temperate latitudes. This is less the case in the tropics, where most of the nutrients in a moist forest ecosystem are in the trees rather than the soil. This potential use of forest land (in medieval Europe probably no less than in Indonesia today) creates a tension between the would-be producers of food crops and those who see the benefits of the *in situ* forests as being greater. In spite of our popular images of the impact of expanding populations, only about 4 per cent of global forest and woodland was converted to other uses in the period 1700–1850.

The virtues of the woodland *per se* are many. Some are intangible: the late Frank Fraser Darling used to say that the Scottish Highland region was so conservative because there were no trees: any form of innovative activity was reported to the neighbours in minutes. Others are easier to document and revolve around the use of the forest as pasture, already mentioned more than once, the use of the woodland usufruct for food (berries, roots, fungi, nuts, honey and wild animals are the main sources) and the wood itself. For some of these uses (especially the latter), many societies developed management techniques to produce particular types of wood and in so doing manipulated the forest away from its natural condition. Cattle like leaves, for example, and so branches normally out of their reach can be cut for them: the practice of shredding was very popular in continental Europe and can be seen in many paintings of the Renaissance and early modern times: Uccello's *Hunt in the Forest* is one example. To grow the juicy and leafy shoots out of reach of cattle a tree such as a willow or hornbeam would be decapitated at a height beyond the reach of the mouth of the largest individual: pollarding is the result. Perhaps the most widespread of all techniques was that of coppicing. A wide variety of deciduous species respond to being cut at the base by putting out shoots quickly and prolifically. These poles can then be allowed to grow to whatever length is needed for particular purposes: for fuel, for implements, for fencing, for wattle-making. Some trees are often allowed to grow to full stature for the larger timbers needed for building. Such forest uses are sustainable over long periods of time and were especially crucial in fuelling industries before the advent of coal: in iron-smelting and tanning for example. Before the industrial revolution, trees were also the source of important maritime materials such as resins and timber for ship construction. Masts often came from forests, since long straight spars are needed; hull timbers, however, often come from park and hedgerow timber, since it is there that trees can spread into the particular shapes needed. But whole areas of woodland may be dedicated to preservation for shipbuilding: the Venetian Republic managed large areas of woodland in the Veneto to make sure that wood was always available for galley construction in the Arsenal.

The margins of the sea were also open to change in this era. There are numerous examples from round the world of reclamation of salt marshes and mud flats, often in sheltered estuaries. The North Sea nations and China perhaps led the way. Such land has the great advantage that it is flat and free from stones, but is subject to inundation at high tides and the soils are salty. So some years have to elapse before anything but grazing can take place and effort has to be expended in building and maintaining a wall to keep out the sea. Centralized political control is an advantage here, since piecemeal efforts are not very viable, unlike the conversion of, say, a piece of woodland or heath, where nibbling is more feasible. Smaller-scale conversions to salt pans, where solar heat evaporated sea-water to leave that most valuable commodity of pre-refrigeration centuries, was a popular transformation of shorelines.

Transformation of estuarine marshes robs the sea of nurseries of fish but there is no evidence of massive impact in the pre-industrial millennia. It is, though, the case that in Oceania many communities living on islands adopted conservation practices, with closed seasons, time-specific food avoidances, taboos and 'conservation officers', so we may infer that fish populations in these places were capable of noticeable diminution. This was certainly the case of those seal populations whose stocks were raided for fur. The case of South Georgia has already been mentioned; another widely quoted example is the Pribilov fur seal fishery (*sic*) in the Bering Sea. Exploitation started in 1799 and perhaps 40,000 seals a year were taken until 1834, when the herd had dwindled considerably, and eventually a stock of 2.0–2.5 million was reduced to 300,000 by 1911; the seals are now managed by an international convention between the USSR, Canada and the USA, and number about 7.5 million.

The designation of a phase of human history as 'industrial' does not of course mean that there was no activity before then which could be so described. Small-scale metalworking has a history almost as old as agriculture but is perhaps more craft-size than industry. However, in China and in medieval and early modern Europe the demand for iron led to large-scale quarrying and mining of ironstones (and indeed of other metals), and to the control of watercourses in order to harness the power of falling water to crush ore and power the bellows, which had to be blown continuously in the blast furnaces found in Shanxi around AD 1000 and common after the fourteenth century in Europe. Many a pond deep in the woods of parts of Germany, France, Czechoslovakia and southern England (where the name 'Hammer Pond' often persists) was once a reservoir for such activities. The greatest influence of such industries, however, has already been hinted at: the need to manage woodland to supply charcoal for smelting. An eighteenth-century blast furnace (the most efficient of its day) using wood from hardwood coppice would have needed the produce of 1600 hectares per year. To keep up a sustainable supply would have required 24,000 hectares of land to be maintained as coppice. At a national scale, early eighteenth-century England's ironmaking needed about 11,000 square kilometres of coppice, equivalent to the area of a square bounded roughly by London, Swindon, Birmingham and Peterborough. This meant that furnaces had to be at least 8–10 kilometres apart since it was cheaper to transport ore than charcoal. Small wonder that some economic historians have looked for the origins of the industrial revolution in a shortage of wood, though this single-factor explanation has been challenged.

Industry is a term which can reasonably be employed to the quarrying of stone, usually for building purposes. Ever since the riverine civilizations developed the cultural notion that the biggest building said what your society was all about, many human groups built high, large and lasting, which meant great quantities of stone. Mount Pentelius bore a gleaming white scar on its side where the marble had been removed to build parts of Classical Athens, and we know that a Cistercian monastery might need 40,000 cartloads of stone in its construction. In terms of environmental change, quarries have been among the most permanent features of a landscape until recently when they have been filled in with municipal waste; even so, the sources of the stone for much of the building of Durham cathedral between 1090 and 1130 are still visible today. Underground mining of salt, too, was widespread whenever a suitable deposit could be found. The wealth of the Archbishop of Salzburg, for instance, depended on the revenues from the Salzkammergut mines – and he could probably have afforded to keep Mozart on the payroll had he so wished.

Naturally enough, industries produced manufacturing wastes: heaps of reject stone and metal slags are often the clues to the sites of pre-nineteenth-century industries. What is not usually detectable now is the poisoning of air and water which must have accompanied these workings: the impact of lead smelting, for example, must have killed every living organism for several kilometres downstream from the furnaces, and the plume of poisonous air will also have had considerable repercussions on plants, animals and human health; in several parts of Europe, distinct subspecies of plants have evolved which can live on soils contaminated by high concentrations of relict metals. If the industry was urban-based then these problems were made worse: the tanning industries of Dutch towns needed special drains to carry away the waste products of leather preparation: they were called *stinkerds*, which needs no translation.

Getting a living in pre-industrial societies could not have been enjoyable for most people. But for those towards the top of social hierarchies, pleasure was important and this is reflected in some uses and metamorphoses of land. Possibly the most important activity for such groups, in many cultures, was hunting. At least 90 per cent of human evolutionary history has been spent as hunter-gatherers, so there is ample opportunity to interpret the killing of animals (and perhaps some October blackberrying) as satisfying very deep-seated characteristics.

Hunting takes two basic forms. The first is the pursuit of wild populations, using netting, traps, falconry and dogs as well as mounted archers and spearmen; the second is to enclose the game species in parks, the better to control their population levels and the type of chase. Both are of considerable antiquity. The first was common in the riverine civilizations and seamlessly extended into imperial domains. As has been common down the ages, certain species were reserved for particular groups or even individuals: in Assyria, the lion hunt was the prerogative of the king, as was usually the red deer stag in medieval western Europe. Hunting of wild populations could reduce them to low levels or they might be diminished by habitat change, such as the conversion of steppe to wheatland for the Roman empire in North Africa. Management of populations might then be undertaken: if the hunted species is a herbivore then killing its predators was popular, as was captive breeding and indeed keeping a few lions in cages in case the king wished to go out hunting. The notion of parks probably comes from ancient Mesopotamia and Persia, where records suggest that gazelle, boar, deer, lions and tigers were kept in large enclosures, though it is not recorded who laid down with whom. In about 300 BC the Maruyan emperors of India created royal hunting forests surrounded by a moat and stocked with tigers, elephants and buffalo, which were to be deprived of claws and teeth; the royal hunters were further protected by bodyguards of armed women, in an early example of equal opportunities. In medieval Europe, kings and princes were responsible for the designation of large areas of terrain as royal forests, which were subject to special laws aimed at protecting the game: in the eleventh century at least 11,000 hectares were subject to these laws in England alone. They restricted the cultivation and pasturage rights of commoners and so were often islands of near- and semi-natural ecosystems, as indeed are some of their descendants today, such as the New Forest in southern England. Kings in need of cash could usually raise some by selling forest rights to the commoners and so cultivation could increase in area after this legal deforestation. Just as the king might have rights to the biggest animal, rights of land management and species availability might cascade downwards to other aristocrats: for example, bishops were allowed rabbits.

The hunt may at times have had utility value in a yield of meat. If this came from a park then the ensemble shared two characteristics with another source of enjoyment: the garden. The two attributes of the garden are that it is enclosed (and near a settlement or dwelling) and that it generates both utility and pleasure. Grass and flowers are usually found, but so are vegetables, fruit and herbs. Only in the tiniest examples, like the scraps of land available in Japanese cities for instance, is the aesthetic purpose totally dominant. Size is scarcely relevant to purpose here: the urban window box may fit the definition, as may the great park at Versailles with 6000 hectares. We go back to ancient Persia (at least) for an enclosed pleasure-park called a *pairidaeza*, which became *paradesios* in Greek, and this fits well the myth of humans being cast out of a paradisical garden (noted for its varieties of trees) to earn their living at subsistence agriculture. Gardens figure in Egyptian wall-paintings, in medieval European books of hours and frescoes, in Chinese and Japanese art and literature and on the ground: the Muslim garden with its emphasis on shade and running water reaches its height in the gardens of the Alhambra in Granada and the Shalimar in Lahore. The name of Plato's garden in Athens, where he taught, was *academe*; institutions following in that tradition now seem to have been deprived of necessary organic fertilizers.

A garden is usually a more or less complete transformation of nature, with imported species probably as important as native varieties. The construction of beds, paths, ponds and plantings makes it possible to impose cultural preconceptions on this restricted area of land: the energy inputs per unit area can be prodigious. So in the socially formal and stratified world of Europe before the nineteenth century, formal and rigidly divided gardens were common around the great houses. Versailles (commissioned from André Le Nôtre in 1662 and taking fifty years to complete) is the outstanding example, with its radiating avenues, compartmented woods and linear canals. Versailles, though, contains a later development of a more informal type. Marie Antoinette had read Rousseau and so had a farm and villages constructed in the grounds of the Petit Trianon to practise a make-believe rustic simplicity. This 1770s development echoes the English Garden made famous by Kent, Repton and 'Capability' Brown, where large landscape parks were planted with informal clumps and groves and watercourses were made sinuous. In the British Isles, the rustic mill of Marie Antoinette's *hameau* was replaced by allusions to the classics: colonnaded temples of Flora or the cells of wise hermits were shown to visitors. This informal landscape garden (which might have as adjunct a walled garden for produce) seems to allow the resurgence of nature (albeit under control) at a time when it must have seemed that the triumph, actual or potential, of mankind over nature was about to be realized. It would be fascinating to have some record of the comments of the Revd Thomas Robert Malthus on a visit (if he ever made one) to Blenheim or Stowe.

If these gardens are a signal that nature is after all to be allowed a place, then we might reflect that some attempts at nature conservation in the mode in which we now know it have a long history. Hunting can quickly deplete wildlife populations and so some monarchs moved to protect species. The Buddhist King As'oka of India, for example, issued decrees (*c*.247–242 BC) listing animals it was forbidden to kill and elephant forests (with a Superintendent of Elephants) were set up. In AD 1107, the Sung emperor in China issued an imperial edict prohibiting the killing of kingfishers for the adornment of ladies of fashion. The survival of deer in England is probably owed to the deer park, with a bank and ditch and a fence of oak stakes too high for the deer to leap over. In the Middle Ages the Bishop of Durham had twenty such parks, out of a national total of perhaps 1900. The combination of the survival of a wild species and its use for sport has continued to the present, for instance in the confident assertions of wildfowlers and deer hunters that they are the true conservationists. However, binding national legislation for the protection of species generally had to wait until the nineteenth century.

We might mention here an enterprise which is not usually undertaken for pleasure (though we might be forgiven for doubts on seeing some present-day commanders on television), that of warfare. The ecological effects of pre-industrial wars were usually temporary, though at the time quite destructive when in places as far apart as Greece and Scotland forests were burned to deny cover to an enemy. When Rome defeated Carthage, fields were sown with salt and wells poisoned; at Massalia, by contrast, Plutarch in the second century AD records that so many Teutons were killed that the soil 'produced an exceeding great harvest in after years'; but it was a spatially restricted method of keeping up the nutrient levels. So although wars could be destructive, the major remnants of those battlefields exist in the heaps of the former bloomeries and blast furnaces that supplied weapons and armour, and perhaps in the ghosts of the angrily slain.

So obvious as to be easily overlooked is the fact that the city itself represents an almost total substitution of one set of ecosystems for another. The pre-industrial city never attained the size and power of its successors, but nevertheless took in energy and matter and transformed them into structures, products and wastes: one result of this is nearly always contamination of the lower layer of the atmosphere with particulates such as dust and soot. In the troubled years of European wars, the functional area of the city itself was extended by a free-fire cleared zone called a *glacis* outside the often substantial walls.

If these are the main transformations to be found in the empires' core and middle zones, are there any differences in the peripheries? Some peripheries were aquired simply as territory, forgotten once the flag had been planted. Others, by contrast, were heavily exploited on behalf of the 'mother country'. As with many other processes, small-scale precursors of larger industrial flows were present, for instance in the form of plantations of cash crops for export such as cocoa, tea, cotton and tobacco. These extended the ecology and social conditions of European farming to many other zones, some of which were unsuited for such intensive cropping and responded with spectacular rates of erosion. In the case of the American colonies of Britain, one of the influences upon westward movement was the rate at which soils became exhausted and eroded under heavy cropping of cotton and tobacco. As a consequence the wooded 'wilderness' to the west had to be conceptualized as a place of savagery, natural and human, so that it could be 'civilized', even if much of it was to go up in smoke. Some peripheries, however, gained status by reason of their production of specialized items: central Scandinavia, Russia and America for ship's timbers and furs in the eighteenth century; east Africa for ivory; and, naturally, anywhere where there was gold and, to a lesser extent, silver.

The period of the empires is, somewhat arbitrarily, terminated here at AD 1800. Beyond that date, however, not all the units which can be thus described made an immediate and full transition into industrialism. China, for example, is still ascribed in some classifications to the category of 'developing countries'; Thailand has inherited the core but not the periphery of the Siamese empire. And before that date, some areas (the case of England is manifest) were already pushing forward towards a fossil-fuel-based economy in which the location of coalfields would be one of the prime determinants of the most thorough environmental transformations that the world had yet known.

Industrialism: the North Atlantic

In AD 1800 the world population was about 957 million, of which perhaps 2 per cent lived in cities. In 1985 it was 4853 million, that is, it had multiplied five times; and nearly half the 1985 population lived in urban areas. No growth of that magnitude could occur without producing great ecological changes even if there were no technological shifts, yet as we know there were immense and far-reaching (in the most literal of senses) developments in science and technology. These conferred on the industrial nations the power to alter their immediate surroundings with a multifariousness that defies comprehensive description and to reach out into peripheral zones with confidence and speed. The empires

Figure 1.3 Early industrialization was small-scale even in its heartland. This colliery and canal were typical of the many scattered developments that pioneered the environmental changes of the nineteenth century. The site is now preserved as part of an industrial museum.
(*Photograph*: I. G. Simmons.)

of the nineteenth and twentieth centuries were held together with steamship and telegraph in a way that would have delighted the Spaniards, who had needed months to get orders to their colonies beyond the Andes. Possibly the one totally new experience of mankind in the twentieth century has been travel at high speed. In India in the 1860s, the railway was held to be the harbinger of no less than theological change: thirty miles an hour would be fatal to the slow deities of paganism, according to one hopeful missionary.

Many technological inventions have contributed to this era, none perhaps less than the clock, but the whole crystallizes around the discovery, transport and utilization of the hydrocarbon fuels: coal, oil and natural gas. Their chief virtue lies in their geological origin, since time has compressed them to the point where the energy content per unit volume is high compared with the newer products of the sun like wood (table 1.1). Charcoal provides the

Table 1.1 Energy content of fuels

Energy source	Content: MJ/kg
Hardwoods	17–20
Softwoods	19–21
Straw	17–18
Charcoal	30
Coal	29
Crude oil	42–44
Natural gas	30–45
^{235}Uranium	80×10^6

overlap but the ratio of dry wood charge to charcoal output in most conversions is about 5 : 1. Thus in developing countries with a high demand for charcoal (especially in urban areas) the ecological impact is especially high, with great arcs around the towns and cities denuded of trees and shrubs.

This concentration is reflected in the steady increase of power output by technological devices. A steam engine of AD 1800 might rate at 8–16 kilowatts, which was improved to 3 megawatts by the end of the century. Steam turbines then took over for higher outputs of 1 gigawatt. In our time, railway diesels top 3.5 megawatts and each engine of a Boeing 747 rates at 60 megawatts; the Saturn C5 rocket rates at 2.6 gigawatts. (A man viewed as a machine in this context might manage 100 watts over a number of hours.)[7] The upshot is a combination of increasing population and greater per capita consumption, which results in faster rates of accumulated energy use. In 1870, the cumulative industrial energy use since 1850 was 3 terawatts per year; in 1970 the equivalent figure was 200 terawatts per year, and in 1986 it was 328 terawatts per year. Most of that energy has been applied to getting and using resources or in travel and communications, some of which carries ideas about how to use even more resources.

One price for the control of all this power is the devotion of land to the extraction, conversion and transmission of energy. Global

[7] The relevant metric prefixes are: k = 10^3; M = 10^6; G = 10^9; T (tera) = 10^{12}.

oil extraction takes up some 1000 square kilometres of the earth's surface and all fossil fuels take up 185,000 square kilometres, about half the land area of Japan. This is just one of the environmental transformations brought about by the processing of energy itself, let alone the uses to which it is put in the shape of machinery. But each of these nodes of energy conversion produces wastes: a coal mine pushes up large volumes of reject rock; a power station burning coal has big heaps of fly ash; if burning oil, then sulphur is emitted into the air; oil refineries exert a high chemical oxygen demand if some of their organic molecules are led off into rivers or the sea. Nearly all energy processing at some stage requires large amounts of water: normally nowadays this is for cooling, which means that a proportion of the energy is being lost as low-grade heat which has done no work. Of the 100 units of energy present in the heap of coal in a power station yard, only 27 are delivered as electricity to the end user and 50 units of the loss are ejected via the cooling water.

The core transformation of this new scale of development is the city itself, along with associated industrial areas. In some ways, we can think of the city as having a metabolism like a natural ecosystem: materials are drawn in, spun into complex structures, some of which act as storage pools, and heat and other wastes are emitted. In constituting this source of demand, the city acts as a potent transformer of the environment both close to its (ever-changing) boundary and then also in a middle zone and on peripheries that may be thousands of kilometres away. The city itself, though, is a prime example of an obliterating influence. To give one simple example, most soils in the densely built-up parts of the urban-industrial area are covered over with impervious materials like brick, tile, setts, slates, asphalt and tarmacadam. Precipitation runs very quickly off these surfaces and by contrast there is relatively little infiltration. The obvious result is a quicker and higher flood peak in the receiving water-courses. At times of normal and predictable intensities of rainfall or snowmelt this causes few problems, but flooding may occur in the city, after an especially heavy thunderstorm for instance. This leads to more environmental modification as the river is straightened and its banks smoothed so as to lead off flood volumes the more quickly. If flooding is relatively frequent in a particular reach, then there will be little investment and with luck the land may be saved as open space.

City building reduces biomass but does not extinguish life altogether. Some species are in fact favoured by urban structures and life. The starling is an obvious example: large numbers find the relative warmth of the city a good place to find roosts after spending the day foraging in fields outside. Sea-birds like the kittiwake will colonize large warehouses as nesting places and expand their populations quite markedly; seagulls are common inland as feeders upon urban waste tips. Rats and mice are well-known concomitants of the urban condition, along with their ectoparasites which may include fleas that spread diseases between humans. Certain arthropods flourish in the relatively warm conditions of not-too-clean housing, under wallpaper and in bedding, for example. Badgers and foxes, racoons and skunks can usually find both food and breeding sites within suburban areas of western cities, and domestic dogs if neglected can form feral packs. Escapes of pets are frequent, but few species manage to make up breeding populations: exceptions may be tropical fish in the warm water downstream from urban power stations, if the water is not too contaminated. Stories abound of massive predators in city sewers, where they have flourished greatly after being flushed away. However, few species mutate to heroic status.

The city needs to eject its wastes and these are plentiful, even before recent trends to a throwaway society in the west. Almost any of these wastes have the potential to alter the environment: in the air, fine particulate matter downwind is thought to increase the amount of rainfall by providing nuclei for raindrop formation; sulphur rains out as an acid liquid to affect most materials, but especially those of buildings on land; holes must be found for municipal solid wastes and thereafter managed so as not to burn nor to harbour excessive rat populations; in freshwater or near the sea, untreated sewage is not only unaesthetic but also causes algal blooms which absorb most of the dissolved oxygen in freshwater and bring about heavy mortality in fish; toxic wastes of all kinds from oil to chemicals will kill birds and other organisms, especially sessile marine creatures like mussels and limpets. If shoreline devices like groynes are erected to stop sand filling up a harbour mouth, for example, then foreshores downdrift which are deprived of the input will start to erode much more quickly: what is gained at one resort's roundabouts is undermined at their neighbour's swings.

In general, the city retains very little of its wastes within itself: like a living organism, it would poison itself if it kept them in-house. A notable exception was Paris in the 1860s, when the city's transport system was powered by about 96,000 horses. Their dung was not wasted: it was applied to 7800 hectares of gardens at a rate of 30 centimetres depth per year and the heat of fermentation was kept in with straw mats and cloches to allow gardeners to produce three to six crops per year of salads, providing 2.4 per cent of the city's protein and adding 6 per cent to the waste heat production of the city. No wonder the artistic life was so creative.

Whatever is said about the city applies with even greater intensity to industrial areas which form part of conurbations or even less densely spread urban conglomerations. Until cheap public transport was available, workers lived near the factories, mines or other installations and created the usual transformations of land, air and water. The lesson of the nineteenth and twentieth centuries is of increasing diversity of materials being winnowed from the environment and a corresponding assortment being led off back into nature. Energy sources and common metals, for example, have been supplemented by a chemical industry (of which fertilizers were the first large-scale industrial product, bringing extra loads of nitrogen and phosphorus to rivers, groundwater and the sea), and a petrochemical industry with many pharmaceuticals and plastics. Their effect on the environment may be indirect, as with pharmaceuticals which combat disease and exacerbate population growth, or direct as in the contribution of plastics to ocean wastes: walk along any strandline, even in the remotest part of the Pacific, and plastics will be the most plentiful item in the debris.

Any summary of the urban-industrial scene and its environmental relations in the nineteenth and twentieth centuries would conclude that two ecological processes have been dominant. The first is the gathering of natural materials in concentrations which had been unknown either in nature or at earlier stages of economic evolution. Nitrogen is a good example: it cycles through all natural systems and is often a limiting factor in the ecology of some populations, especially in water. In pre-industrial times, it was often scarce in agricultural systems and was thus 'hoarded' by recycling organic matter onto crops as much as possible. With industrialization, large quantities are fixed from the air and applied to the land at the same time as city growth means that a lot of sewage is being put into watercourses. Hence, eutrophication is common and many aquatic systems are no longer limited in nitrogen, so explosive growth of algae, for example, can occur. These then decay under bacterial action, which absorbs oxygen and results in the killing of fish.

The second process is the creation of molecules which are unknown in nature, as with many industrial chemicals and environmentally persistent materials such as pesticides. Generally speaking, such synthates have no breakdown pathways in nature (or at best only very slow ones) and so remain available for intake by organisms. Accumulation within biota can occur, with both lethal and sub-lethal results: in wild birds, eggshell thinning is a common response to high levels of certain pesticides. Other

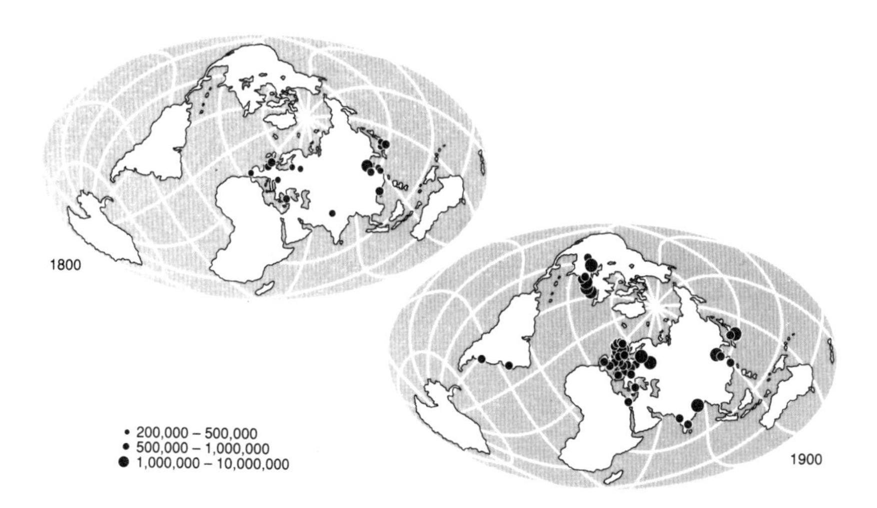

Figure 1.4 World cities: growth in number and size 1800–1900.
(*Source*: B. L. Berry, 'Population', in B. L. Turner et al. (eds), *The Earth as Transformed by Human Action. Global and regional changes over the past 300 years*, Cambridge University Press, Cambridge, 1990, figs 7.3 and 7.7.)

Figure 1.5 The apogee of the industrial and the precursor of the postindustrial is the world city: a concentration of energy and materials itself, with the power to move capital in order to facilitate the manipulation of the environment almost anywhere. This is Toronto, but it could be one of many such cities.
(*Photograph*: I. G. Simmons.)

compounds may react with substances which are critical in the natural environment: the stratospheric ozone layer protects life forms from an excess of ultra-violet (UV) light, but is attentuated by reaction with chlorofluorocarbons (CFCs) to simple oxygen, which does not filter ultra-violet light.

If these two ecological processes are dominant under industrialism, then they are consequences of two other trends which we have noticed before: (1) the intensification of processes, and (2) their spatial extension. Intensification in this era means the application of more energy per unit area in the course of all kinds of economic and cultural processes: in the nineteenth century not only did the absolute amount of energy mobilized by humans go up, but the per capita usage as well (table 1.2). Extension means the reaching out of societies from core areas to exert political control as well as economic dominance, and was made much more effective by the steamship (especially when carrying troops) and the telegraph. Thus the consequences of intensification and extension of industrial systems was seen not only and obviously in their core areas but on the peripheries much more than in preceding periods. Those core areas in 1800 reflected the transition to the Atlantic zone's dominance, but included no North American cities of world status; whereas by 1900 most world cities lay between Chicago and Vienna, with outliers in South America, the Indian subcontinent, China and Japan; and by 1985 they fell mostly in an arc from Los Angeles to Beirut and then eastwards, outlining the coast of Asia as far as Tokyo, with outliers in South

Table 1.2 Energy input and population density over time

	Energy input (GJ/ha)	Food harvest (GJ/ha)	Population density (Persons/km^2)
Hunter-gatherers	0.001	0.003–0.006	0.01–0.9
Pastoralism	0.01	0.03–0.05	0.8–2.7
Shifting agriculture	0.04–1.5	10–25	10–60
Traditional farming	0.5–2.0	10–35	100–950
Modern agriculture	5.0–60	29–100	800–2000

Source: V. Smil, *General Energetics*, Wiley, Chichester, 1991, p. 239.

America, South Africa and eastern Australia. Nevertheless, the gamut of human activities to which these intensifications and extensions have been applied is not radically different from before: food-getting is still basic, pleasure still sought, and warfare is still allowed to plague us.

Even in the hi-tech, hi-sci societies of the West, people still have to eat; and indeed some of them have elevated simple nutrition into a major art form, just as others have degraded it into a massively picky set of fads. Industrialization of food production has meant in ecological terms the addition of much more energy to the system to supplement the solar content. From the hunting-gathering stage to the present, the energy inputs have risen by four or five orders of magnitude and the number of people supported likewise. The application of industrial ways to food production means that it is now better to talk of a food system, since the whole flow from fertilizer factory to dining table is, in western economies, tied into the potentials and imperatives of the industrial lifestyle. Some of these stages have a distinct role in environmental change, but others (for instance, flour milling) are not distinguishable from other industrial processes.

An industrialized food system may still centre on a field producing a crop of plants or animals, though there are many other production sites such as indoor rearing sheds, aquaculture installations, concrete-floored outdoor pens, cellars for mushrooms and oil refinery-type towers for mycoproteins. Upstream from these sites, the production of machinery and chemicals is important; and downstream there are all kinds of drying, processing, transporting, storing, wholesaling, retailing and preparation to be carried out. For our purposes we need to note:

1. Energy derived from fossil fuels is needed for most of these. Some food becomes heavily energy-negative by the time it reaches the mouth: we do not eat lettuce from heated greenhouses in January for its nutritive content. The highest proportion of this energy is expended in the home.
2. Wastes are possible at any stage, but food processing produces large concentrations of organic wastes which need special care if they are not to become noxious.

It is not surprising, then, that in the field one of the main effects of modern intensive agriculture is the simplification of ecosystems. The battery of tools for eliminating competing organisms is now enormous and highly effective, and in getting rid of weeds and pests can also remove much neutral or even positive-value wildlife. The use of heavy machinery makes large fields an economic necessity (increasing the chances of sheet erosion and windblow), and heavy capitalization may make production levels so important that irrigation is worth while even in areas normally marginal for its use.

In arid zones, the commonest form of intensification has indeed been irrigation. In the Middle East, for instance, 70 per cent of agricultural production comes from the 35 per cent of cultivated land which is irrigated. In the UK agriculture consumes about 3 per cent of total water withdrawals; in India and Mexico the figure is about 90 per cent and in the USA 42 per cent. The provision of all that water means a great deal of dam construction, well drilling, pumping and other land use management, including the regulation of forest which might otherwise transpire a lot of water. Reduced to energy terms, the energy input for irrigation in the tropics usually exceeds that for fertilizers.

Extension of the cultivated area of the world in this period has been of two main types. The first is the breaking of the grasslands after 1870, in which many of the world's natural grasslands were converted to grain production to feed the burgeoning populations of core regions: areas of North America, Australia and central Europe feature prominently in the 281 million hectares of mid-latitude temperate grasslands cultivated for the first time, mostly for wheat, in the nineteenth century (table 1.3). The second is irrigation, used to make cultivation possible at all, rather than for intensifying the use of already tilled areas.

Soils have borne the brunt of these changes: they have been subject as never before to accelerated erosion (the examples of the United States' Dust Bowl and of Khazakstan in the 1950s are frequently cited), and to subsequent coastal modifications: it is an item of conventional wisdom that much of Nepal is relocating itself in the Bay of Bengal. Soils are also prone to the age-old problems associated with irrigation, such as salinification and waterlogging. In Syria, the Punjab and Iraq, 50–80 per cent of all

Table 1.3 Percentage changes in global land use 1700–1980

	1700–1850	1850–1920	1920–1950	1950–1980	1700–1980
Forests and woodlands	−4.0	−4.8	−5.1	−6.2	−18.7
Grassland and pasture	−0.3	−1.3	0.5	0.1	−1.0
Croplands	102.6	70.0	28.1	28.3	466.4

Source: J. F. Richards, 'Land transformation', in B. L. Turner et al. (eds), *The Earth as Transformed by Human Action. Global and regional changes over the past 300 years*, Cambridge University Press, Cambridge, 1990, p. 164.

irrigated land suffer from these effects. Equally, the consequences of irrigation projects for public health are often adverse: at least ten human diseases are associated with them.

We can conclude that the world's food production systems have performed very well in the Atlantic period, especially for the world of Europe and North America, who have had access to much greater quantities of protein, raw materials, fertilizer and energy and have woven these into dominance over the world's economy and parts of its ecology as well. This must have been so since at least AD 1600, but if we need a symbol then the internal combustion engine in the form of the tractor is probably the best.

One hallmark of the nineteenth century and after has been the emphasis on production of all kinds and this has created some environmental change in systems whose rates of production was traditionally rather leisurely, such as forests. The impact of the twentieth century in particular has been to treat them more like food crops, with a shorter cropping interval and more intensive management of the crop with fertilizers, pesticides and mechanized harvesting. Single-species forests are easier to manage, although more prone to pest and disease outbreaks. The contribution of industrialism has extended to the intercontinental transport of tree species deemed to be of commercial potential (from the 1840s North America's west coast was a particularly important source, as later was Australia). Large-scale money-making plantations have been extended to the tropics and some companies will now trumpet their tropical reafforestation policies without quite mentioning that a forest which originally had 200 species per hectare is being replaced with one pine or eucalypt species.

Some diversification of this pattern has been introduced by the realization that forests have many other values. Watershed protection, wildlife conservation and outdoor recreation are the most prominent of these, and management of the forest environment will vary according to the mix or mosaic of uses. Recreation, for example, will probably mean forgoing timber harvests in some places to protect valued campsites or picnic areas or favourite scenic views, or making sure that post-logging silt is kept out of the streams during the summer at least.

Forestry has long been a favoured use of the periphery. From early use of Swedish and Russian forests for ships' masts through colonial administrations' forest services (Java in 1929, Burma in 1855, for instance) to the large-scale buying of Southeast Asian timber by the Japanese, it has been an extensive land use which could be controlled at a distance (apart from felling, nothing happens very quickly) and often with a distinct feeling of superiority: the ascendancy of long-term timber production policies over the 'waste' of shifting cultivation is a good instance. Only very recently has a role for certain forests in the tropical lowlands in the ecology of global climate been recognized at a scientific level, and that knowledge has not yet translated into very much change in environmental manipulation on the ground.

If we want an example of the one natural flow that has undergone most transformation at human hands through time, and in the present era no less so, then it must be water. Already mentioned more than once in this section, it is perhaps worth the risk of being repetitive to emphasize the outreach for it which an industrial society will undertake. To ensure high quality and reliable supplies for an urban-industrial area is a considerable undertaking, and in the nineteenth century the competition for the catchment area on the wet hills of the Pennines above the industrial zones of Lancashire and Yorkshire was something of a microcosm of the imperial slicing-up of Africa in which their rulers were then engaging. This century has seen much bigger pumping schemes and inter-basin transfers, and even a proposal in North America to take water from the Yukon River and lead it, eventually, into the shower heads and shampoo parlours of Los Angeles. Desalination is a reality in a few parts of the world (those with cheap energy or no choice: contrast Abu Dhabi and Malta respectively), and grandiose plans to tow icebergs from Antarctica to Saudi Arabia or Western Australia surface from time to time.

One result of energy surpluses in the industrialized world has been an increased appetite for pleasure: people have more time and disposable income, and transport makes shorter breaks possible from more serious occupations. In Britain, a weekend at the Metropole could only have post-dated the railway line to Brighton. The interaction of steam power and pleasure is seen in several environmental changes. At the demotic end of the market is the growth of the nineteenth-century seaside resort made possible by the railway. Regularization of large stretches of coastline followed, with consequent effects upon sediment supply and deposition, so that some resorts' building programmes cut themselves off from their supply of golden sands, while others encouraged the laying down of good quality blackish muds. Sewage pipes leading out as far as low tide mark were apt to do service in lieu of actual treatment (they still do) and thus encourage the concentration of scavenging species such as gulls. Further up the social scale, the railway made feasible the August break of City financiers to the grouse moors of the north and, coupled with the breech-loading shot-gun and plentiful casual labour, projected grouse-shooting into prominence as a late summer-to-autumn occupation for the well-off. Management of grouse moors has meant the production of a heather-grouse monoculture, with fire as a principal tool. Such monocultures are notoriously unstable and so wildfire produces severe soil erosion. Eruptions of the pests of both heather and birds have caused anxiety since the early twentieth century in north-eastern England and Scotland.

The harnessing of steam-powered transport to the purposes of outdoor pleasure among the rich has a thickly-textured skein of history. In Trollope's *Palliser* novels, for example, well-known members of London society have little trouble in slipping up to famous Midland packs for a day's fox-hunting. Fox-hunting, whether in England or Virginia, in fact requires a pre-industrial landscape in which there are plenty of small woodland copses and fields are fenced with wooden posts and rails or bounded by hedges; neither wire fencing nor modern 'prairie'-sized fields are helpful. But in keeping up the expensive pack and its servants and in getting, before motoring became popular, the right people to the hunt on the right days of the week (and of course Sunday was not available), industrial money was a helpful adjunct to the older money of the shires. The steamship was also important in taking selected people on safari in Africa or India, to shoot big game. This practice followed for years the tone set by royalty and the plentiful kills recorded in the *London Illustrated News* and similar organs suggests that some populations of leopard, tiger and elephant must in places have come under stress in about the 1920s. But the social cachet continued in Britain until an outbreak of diplomatic whitlows on trigger fingers in India in the more conservation-minded 1960s.

No history of pleasures and their environmental relationships can omit outdoor recreation and tourism. Nowadays these are mass activities, but until the 1950s were in general the preserve of elites, so relatively little management was undertaken. Nevertheless, in the 1920s, for instance, the National Parks of the USA were very popular vacation spots and so roads were taken into remote valleys and the whole apparatus of tourist accommodation and related catering was erected, to be further enlarged during the prosperity of the 1950s. Recreations such as skiing were not seen as incompatible with the preservation of nature, so that in Canadian National Parks, such as Jasper and Banff, ski runs and cable cars, hot-spring hotels and whole towns were seen as part of the winter experience; the European Alps in the post-1880 period served as an exemplar. In cities, too, tourism has always had to contend with the disruptive effects of the visitors at the same time as relieving them of as much money as possible. Readers of E. M. Forster's *A Room with a View* (1908) are reminded that the Giotto frescoes in Florence's church of Santa Croce could get uncomfortably crowded even in those more spacious days.

Extinction of species is a normal part of organic evolution, but human activity in modern times has been a potent accelerator of demise. A guess at the 'natural' rate would put it at 1 per cent of the given number of species every 2000–3000 years (though that assumes a rather gradualist view of evolution, which may be more of a series of equilibria punctuated by bursts of differentiation and extinction). Given this, the human effect has been to extirpate species at about 40–400 times the 'natural' rate. If we count birds and mammals, for which the data are reasonably reliable, then one species or subspecies became extinct every four years in the period 1600–1900, and one each year thereafter. This rate seems still to be accelerating, though it may be hoped that feedback will slow it down soon via protection policies. One result has been the transfer of land to conservation uses and its subsequent management for the protection of species, assemblages of species or whole regional ecosystems (or indeed a continent, in the case of Antarctica). The management aims are to reverse the processes of overkill and habitat destruction. This latter is currently the most important and is currently manifested in processes like deforestation, introduced animals like goats and rabbits, and drainage of wetlands. The ploughing of the Great Plains was a major factor in the extinction of the passenger pigeon, just as agricultural changes, including ploughing, have eliminated the great bustard from most of Europe. The Norfolk Island parakeet has come to the brink of extinction due to a disease caught from domestic poultry.

The nature reserve in its more recent forms is the direct descendant of those royally protected areas of pre-industrial times, in the sense that management of the protected populations is deemed to be essential. In some movements there has been during the early twentieth century a very romantic 'hands-off' approach, but it is recognized that only rarely will this work if the reserve is an island of near-natural ecosystems in a sea of intensive land use. But in many nations now there is a hierarchy of designated areas for nature protection (just as there is for human-created monuments), and the important feature to discover for our purposes is the extent to which there is active management rather than a pious hope that a fence and a signboard will ensure the perpetuation of a rare flower or a large predator.

No account of industrialization can pass without discussion of wastes. Just as King Midas managed to turn everything into gold, so human societies can produce garbage from almost anything and the richer the society, the greater the volume and diversity of the garbage: ask archaeologists about their best sources of information. The term pollution is most often applied to these substances (and energy in the form of noise) and this has taken on moral resonance. For environmental history we might want to recognize that scale is important and that it is only with gases in the atmosphere that humans have been able to produce a truly global set of contaminants. An accurate date for the beginning of this phenoemenon is not obtainable, but the departure of carbon dioxide levels from what nature had produced seems to have been in the middle of the nineteenth century. Methane came later, as did artificial gases like the CFCs and other chlorine-containing gases and aerosols. These gases in particular are at the heart of the radiative forcing of global temperature which is the focus of the 'greenhouse effect' to which so much attenion is currently being devoted. Other pollution problems with a high profile are not normally global in their incidence, and some, like 'acid rain', are an interesting echo of the nineteenth century and its heavy fallout of sulphur and a legacy which includes in Britain the vegetation of much of the Pennine moorlands. Here the dominant cotton-grass of the species *Eriophorum* seems to have replaced bog-mosses at much the same time as steam engines replaced water power in the industrial valleys. The legacy also includes onion-skin peeling of stone on many buildings: in the mid-nineteenth century Durham cathedral had a four-inch layer of rotten stone scraped off the entire outside surface of the structure.

No such discussion is complete without some mention of the greatest waste of all, that of warfare. Two conflicts stand out for the environmental alterations they caused: the trench warfare in Flanders during the First World War, and the Second Vietnam War of 1961–75. In the first of these conflicts, the battle-front was a high-energy ecosystem, with a rapid transformation of materials and energy into noise and heat. Diversity was reduced largely to men, horses, lice and rats and the landscape became a vast morass of muddy pools and mires resulting from choked streams. Except where deliberately preserved, however, this landscape reverted in about twenty years to an agricultural set of uses, with the notable and apparently permanent exception of the graveyards. In Vietnam, chemical warfare against vegetation was widely used and herbicides were dropped on *c.*10 per cent of the area of South Vietnam. Much forest was also cleared by mechanical means and as the result of bombing: a single B-52 bomber might produce a cleared swathe of nearly 300 hectares through tropical forest.

Such forest when cleared does not always regenerate: tough grasses may instead form a permanently deflected succession. It appears too that mangroves may not recolonize the silty and muddy areas from which herbicides cleared them, and so coasts are open to much greater storm damage. Beneficiaries from the war were, it seems, malarial mosquitoes, which bred in the many stagnant pools, and tigers, which scavenged on corpses.

It is difficult to sum up the industrial phase of human history whose powerhouse was either side of the North Atlantic. For a start it is not over: the fruits of it are still being transferred to various parts of its periphery. Equally there are parts which seem to be ready to move on to a different kind of economy and with it the potential for different types of environmental linkages. Though clearly not yet history, a brief discussion will avoid leaving this account dangling over the edge of time.

Towards a global economy

The repertoire of energy sources which is so central to environmental transformation acquires in these last years the addition of nuclear energy, generated from the fission of ^{235}uranium. This, however, only enters an economy as electricity and so is often out of the reach of the poorer economies, especially if they can produce electricity from hydropower. A whole slew of 'alternative energies' is currently under development, but their full potential for environmental transformation, either on-site or by virtue of the energy that they harness, is not yet apparent. The other trends that are apparent and which may concern us seem to be:

1. A shift in the centre of gravity of the world economy towards the nations of the Pacific Rim, with California, Japan, Hong Kong and Singapore in the lead.
2. A globalization of world culture under the influence of ubiquitous and instant telecommunications, with TV exerting the most vigour. This is backed up by the instant electronic transfer of capital funds, which seems likely to have strong influences on economic development.

3 A countervailing trend towards the small and the local in which identity is prised away from the global image-makers and developed as a series of non-conforming 'bioregional narratives'. This is especially the work of the adherents of the deeper shades of Green opinion, but the recent upsurge of nationalism may turn out to be part of the same processes.

The future of the natural environment is, however, certainly bound up with human culture since there are so many possibilities physicially open to human societies. One of these is to try to replace nature almost entirely by artificial systems completely under human control: the ultimate outcome of the Renaissance project, as it were. Some progress seems to have been made in this: the average human biomass world-wide is about 1.6 grammes per square metre; densities of 5 grammes per square metre can be sustained without relying heavily upon imported food, but many nations have exceeded that level: China at 5.3, the UK at 12.0, Bangladesh at 33.0 and Hong Kong at 250 are examples. In world terms:

Total organic matter production = 224.5×10^{15} grammes
Direct use by humans = 7.2×10^{15} grammes
Indirect use or conversion by humans = 42.6×10^{15} grammes

Thus about 39 per cent of the production of organic matter on the planet is appropriated by humans.[8]

Another possibility is to revalue the wild, especially in its most pristine form, which is usually labelled *wilderness*. In chapters 2–4 we shall examine the ecology of the replacement processes in both general terms and those of specific environments and places; that complete, the book will end with a look at the cultural programmes within which human-induced environmental change is always set.

[8] J. R. Diamond, 'Human use of world resources', *Nature, Lond.*, 328 (1987), pp. 479–80; C. J. Pennycuick, *Newton Rules Biology* (Oxford University Press, Oxford, 1992).

2

'*Might-be maps*': The Ecology of the Wild and the Tame

Humans live in two worlds: an ecological world in which we have basic metabolic requirements analogous to any other form of life; and a psychological world in which our cognitive faculties have enabled us to construct complex cultures. In our manipulation of the biological and the physical we are much influenced by the cultural (which of course includes technology), just as at various times culture has been influenced by the environmental. The immediate task in this chapter is to bypass the history of ideas about the biosphere (which we will look at in chapter 5) and to concentrate on a more scientific account of how human societies have brought about a metamorphosis of the ecological communities around them. The account will focus on the interaction of soil, water and life in these systems rather than on the rocks beneath or the air above, though they too will need to be mentioned from time to time.

The basic concepts for the explorations to be carried out in chapters 2 and 3 are:

- The science of ecology. Ecology is concerned with the relations between plants and animals and their non-living environments, and in particular with the exchanges of energy and matter which result in the population dynamics of particular species. Ecology has in the past often been concerned with finding equilibrium points in the systems which it studies, whether at a global scale (for instance the equilibrium vegetation of the globe after the disturbances of the Pleistocene glaciations) or a smaller scale, such as the dynamics of plants and animals after a savannah fire. Today, the notions of predictable progress of biological communities towards some stable state are less popular: the idea gaining dominance is of rather more random mosaics of species which never reach any predictable equilibrium but exhibit the kind of open-ended behaviour described by chaos theory.

- The concept, nested within ecology, of an interactive and dynamic interaction of all the components of an ecological system: plants, animals, soils, water, humans, meteorological factors and time: an ecosystem. The concept is scale-free and can thus be used of a drop of water in the crook of a tree branch as well as of the planet Earth as a whole. In modern studies of ecosystems, the cataloguing of species and the mapping of, for example, soils has been much supplemented by measurements of energy flow from the sun through plants and herbivores to carnivores and then to the decomposers of dead organisms. To this is added the circulation of inorganic nutrients such as carbon, nitrogen and phosphorus within a given ecosystem and between ecosystems.

The basic ecological tenet which we bring to a study of human history is, then, that natural ecosystems have received numerous impacts at human hands. Some of these will have produced only temporary biological and physical changes, others have brought about permanent metamorphosis. Hence we must examine the kind of state which we expect to encounter and also discuss the ways in which humans can alter this ecology away from the primeval, though without necessarily producing the paved.

Succession and climax

The ecology of natural ecosystems has for decades rested on the notions of succession and climax. Successions were of two types: those driven by external physical or chemical events such as sea-level change or volcanic eruption (*allogenic* successions) and those in which the biota themselves created the conditions in which their successors of different species might flourish, *autogenic* succession. The basic idea was that simple communities colonized bare areas created as part of allogenic processes. Thus spreads of cooled lava or volcanic ash like the dust and ash deposits of the Mount St Helens eruption of 1980 are now becoming colonized with plants. Other examples are land which is emerging from beneath glaciers or ice-sheets, as can be seen downstream from glacier toes in any Alpine valley, or land being built from the sea or emerging from it as a result of falls in sea-level. Sand-spits being formed along a coast due to increased silt supply exemplify this last process: even when the sand is mobile some species can establish themselves, and they act as forerunners for others. Such early stages of succession are often unstable in the sense that they involve open ground which can get blown about and bury nascent communities, and which is open to erosion and nutrient loss and invasion by exotic species. But it also provides – literally – the seed-beds for later species which can only tolerate more stable habitats; classically these later stages contain a greater variety of species and are more resilient to change on a large scale, though senescence and death of individuals is a constant component of the ecology. The endpoint was thought to be in equilibrium with the zonal climate on a global scale. Its species composition was diverse and the weight of living material per unit area (biomass) was the highest that ecological and geological evolution could produce in the climatic circumstances; that is, the quantity of living material was limited largely by temperature, day length and water availability. This condition was the endpoint of succession and was labelled the climax community, the 'mature' ecosystem. This generally shows itself to be of a stable composition which is independent of the initial species admixture. Several of the contrasting characteristics of early and mature stages of succession are shown in table 2.1.

Though an apparent endpoint, this system had to replace itself since the life-span of individual organisms was limited: few trees, for example, have the 1000–2000 years' longevity of the genus *Sequoia*, the redwoods of California and southernmost Oregon.

Table 2.1 Characteristics of successional stages

Ecosystem attributes	Early stages	Mature stages
Biomass	Low	High
Food chains	Linear, mainly grazing	Weblike, mainly detritus
Inorganic nutrients	Outside biota	Within biota
Species diversity	Low	High
Pattern diversity	Poorly organized	Highly organized
Size of organisms	Small	Large
Life cycles	Short, simple	Long, complex
Mineral cycles	Open	Closed
Mineral exchange between inorganic and organic components	Rapid	Slow
Role of dead matter	Unimportant	Important source of nutrients
Resistance to external perturbation	Low	High

Source: Adapted from E. P. Odum, 'The strategy of ecosystem development', *Science*, 164 (1969), pp. 262–70.
Note: Since the publication of this work, ecological theory has shifted views on some of the characteristics described: ideas on the organization of pattern diversity, for example, have changed. The table is therefore included so as to show the kinds of distinctions that can be made between ecosystems at different stages of their evolution. Odum makes the point, too, that human activity tends to produce 'early stage' systems rather than 'mature' ones.

Classical climax theory held that replacement was a matter of the piecemeal replacement of plants as individuals became old and died, so that in a mature forest an old giant fell and in the subsequent clearing the processes of succession went on until once again another individual, but of the same climax species, stood in its place. Thus in a natural world there were belts of mature vegetation (with associated soil types and animal communities) which had developed since the end of the Pleistocene ice ages (that is, in the last 10,000 years) into self-renewing and resilient natural formations. But there is probably no overall global equilibrium of vegetation in the present interglacial[1] period since there is always the driving force of climatic change.

More recent research has complicated this representation. It now appears, for example, that the direction of succession is not nearly so arrow-like as the classical picture suggests. Disturbance during succession by forces such as fire and windstorms produces a mosaic of patches, each with their own species composition and ecological history. Natural areas are more often now viewed as 'patch-dynamic' landscapes in which there may be, it is true, relatively constant proportions of patch types on large areas through time, but no single location ever becomes stable in the sense of the classical model. There may well be a continuity of climax types which vary along environmental gradients and whose rate of transformation is so slow that we cannot detect it. One immediate consequence of this for our story is immediately apparent: if one of the factors producing patches is lightning-induced fire, but if also human communities are present which clear land for agriculture using fire, then the evidence for the past may make it virtually impossible to disentangle the two causes: it may be possible (indeed, likely) to label a natural patchwork as 'man-made' and vice-versa.

Nevertheless, the notion of a progression through time of vegetation communities towards a relatively stress-tolerant, resilient (that is, tending to resume succession towards a state of higher biomass) and energy-maximizing condition is too consonant with observation to abandon entirely. We see that forest clearings in the tropics, once abandoned by agriculturalists, grow successively more and denser species of woody plants until a few trees once again join the high canopy; in temperate lands, abandoned coppice quickly grows out to a dense woodland of multiple-stemmed trees. In eastern North America, deserted agricultural land has in this century been documented as its bareness is reclaimed by annual plants, perennial herbs, then bushes and finally, for the moment, forests dominated by species which are intolerant of shade. So the ecology of any area of apparently 'wild'

[1] This is an assumption, of course, but there seems no reason to argue that the latest glacial era in Earth's history is finished just because there was a major withdrawal of ice about 10,000 years ago.

ecosystems has to be interrogated in a spatial and historical context to see to what extent it lives up to the model in our present perceptions of what a 'natural' ecosystem should be.

A global mosaic

Given, however, the observation that some parts of the globe are relatively untouched by humans and have had millennia of relatively stable natural conditions in which to form ecosystems, and that by contrast there are areas virtually without life because they have been subject to volcanic activity or recent changes in sea-level, then we can view the world as a mosaic of different kinds of biological systems.

First, there are the mature systems, where the biomass appears to be the greatest that natural conditions will support. The remaining areas of closed-canopy tropical forest, the Boreal coniferous forests and tundras of Eurasia and North America, remnant wild grasslands of central Asia, much of the deep oceans and many higher mountain ranges fall into this category. At the other extreme, there are areas with little life at all: examples of natural occurrences have already been given, and we must add to them examples produced by human activity such as land covered with concrete like an airport runway, or terrain so toxified (for example, by carbon or metal fallout from refineries) that even bacteria find it difficult to survive. A third category is the community in an early stage of succession where there may be a lot of bare ground and either many species which can only tolerate open conditions or perhaps only a few species which can live under rather extreme conditions, but which will eventually if left alone acquire the characteristics of more mature ecosystems. Human societies mimic these systems in their construction of, for example, agricultural fields (a lot of bare ground, often only a few species – and indeed the fewer the better in modern cultivation) and some kinds of garden. Lastly there are mixed systems: a coastal sand dune complex in western Europe may have some bare sand close to dense shrubs or even low woodland; rice cultivation in China may take place in a tessellation of *padi*, woodland, open grassland and areas of grazed scrub; a new forest plantation of, for instance,

Eucalypts may abut on one side a vast new mechanized terrace for growing wine grapes and on the other a dense three-dimensional system of small fields with maize, trellised vines and grassland for cattle, as in the Minho area of northern Portugal.

The point of this classification is to note that if we can imagine an epoch when all these systems had had time to assemble after the upheavals of the Pleistocene ice ages but before human influence became very marked (and 'imagine' is probably the right verb), then there was some kind of balance between all these four types of system. All were trending to the mature but some were constantly thrown back by natural events; inert land was continually being created and then undergoing succession, the outcome of which involved chance rather than an apparently inexorable progression to a particular type of mature ecosystem. The effect of human dominance of many areas of the globe, however, has been to alter this balance: the quantity of the mature has been much reduced and replaced with the successional and the mixed and even the inert. In terms of our present task, we have to chronicle the disappearance of much of the mature at human hands. In this process, we shall need some definitions for changes wrought by human hands. A regional landscape or ecosystem not affected by people can be labelled *natural*, corresponding to the German idea of an *Urlandschaft*. If there is some human influence but the same structure as the orginal ecosystem (for example, a woodland remains a woodland), then *sub-natural* can be used. If the flora and fauna are still wild species (that is, not domesticated) but the structure is different, then *semi-natural* will be employed. This would apply to a moorland or a heath derived from an orginal forest or grassland. Lastly comes the *cultural* ecosystem, in which the dominant species are those of domestication (as with crop plants and animals) or importation, as with North American conifers planted in Great Britain or Eucalypts in India. Terms such as *agroecosystem* are sometimes used for particular cultural vegetation types. The relations of this and other terminologies are given in table 2.2.

Human-induced alteration of natural systems

The ways in which human societies alter the natural world are wonderfully various, and here an attempt will be made to charac-

Table 2.2 Relations of different systems of terminology of human-altered systems

Development stages terminology, after Odum*	Degree of naturalness terminology – no humans	Degree of naturalness terminology – human presence	German historico-cultural terms
Inert	Natural	Cultural	*Kulturlandschaft*
Successional	Natural	Cultural to sub- and semi-natural	*Kulturlandschaft*
Mixed	Natural	Cultural to natural	*Kulturlandschaft*
Mature	Natural	Natural	*Urlandschaft*

* E. P. Odum, 'The strategy of ecosystem development', *Science*, 164 (1969), pp. 262–70.

terize some of the processes. The way in which they are described cuts across conventional economic categories and concentrates instead on the impact of a human action upon the ecosystem concerned. If we assume that we start with a natural system, then these are the ways in which sub-natural, semi-natural and cultural consequences can be produced (see table 2.3).

Deflection is a method of preventing natural succession from going beyond a certain stage because the earlier stage or stages have more value to a human group than any eventual mature ecosystem. It often aims at keeping the vegetation at a sub-natural or semi-natural condition in which the species involved are all wild (rather than tamed or domesticated) but the frequency of them and the physiognomy of the community are kept at a culturally desired state. An example is the use of fire by pastoralists in semi-arid environments like those of the savannahs of Africa. Here, fire is used to burn off dead vegetation at the end of the dry season so that new shoots activated by the rain are unshaded by the old leaves and thus provide a quicker source of food for cattle. Burning also consumes young and tender saplings so that tree growth is discouraged. On the moorlands of England and Wales, traditional (that is, pre-nineteenth-century) burning practices by sheep-farmers did much the same: last year's stems and dry leaves of the purple moor-grass (*Molinia caerulea*) were burned off and on peaty substrates this encouraged the early growth of the next year's leaves or even the replacement of *Molinia* by cotton-sedge (*Eriophorum* spp.), which will start to leaf even earlier in the year. The animals thus benefit from better early-spring nutrition.

Fire is a near-perfect tool for deflections of these kinds, where a different species mix or vegetation type is sought but where only a limited amount of labour input by humans is desired: in effect, the energy required to deflect the development of the ecosystem comes from stored but recently fixed solar energy in the form of dry plant materials. It is thus possible to use it in any ecosystem in which there is enough dry material at some season of the year to burn for more than a few minutes, and there are few which do not fulfil such criteria; even primary lowland tropical forests are said to have burned under certain conditions when the climate was a little drier, and the English mixed-deciduous forests of prehistoric times seem to have been subject to fire from time to time. Very

Table 2.3 Summary of human actions by cultural ecology stages

	Deflection	Simplification	Obliteration	Domestication	Diversification	Conservation
Hunter-gatherers/ early agriculture	Frequent, by use of fire	Rare	Only at a few settlements; some spp. made extinct	Dog by hunter-gatherers; then others	Domesticates add to gene pool	Yes: to avoid over-use of species
Riverine agriculture and pastoralism	Some permanent	Agriculture replacing wild; pastoralism simplifies	At permanent settlements	Widespread	More domesticates	For pleasure
Agriculture-based empires	Permanent and temporary	Increased efficiency in removing competitors; predators exported	Spp. extinction starts on modern scale	Fewer added to repertoire than previous period	Transfer of spp. becomes intercontinental	Pleasure of the rich; some intrinsic value accorded to wild
Industrialism	Much permanent change	Widespread as cash crops replace wild and traditional systems	Obliteration through chemical as well as physical means; very widespread	Science of genetics enhances predictability	More transfers; new varieties	Value of wild and traditional becoming realized: all kinds of conservation
Global economy	Not different from above	Continues though when choice exists, rate slows	Extinction rate still high though increasing attention paid to it	Genetic manipulation holds much promise	Many organisms transported and capable of mutation	New holism of humanity and nature?

many ecosystems with a dry period, of course, are subject to natural fire started by lightning and so we surmise that early human communities imitated the nature whose benefits were otherwise more random in occurrence. Of the various types of fire found naturally, the surface fire, confined to dry matter at ground level, is the most useful as a manipulative tool.

The benefits of fire to humans are usually realized by repeated burning of a patch of terrain. A one-off burn may yield some extra food from animals caught fleeing the fire and smoke, but plant 'crops' are more likely to be produced after a few years' treatment. The deflection of the ecology of the area proceeds in a number of ways once fire is repeated annually or even more frequently. For example, fire-resistant species may come to dominate the vegetation. Such plants may have an adult phase with thick bark which protects the trunk and branches of woody vegetation; the Giant Sequoia of the Sierra Nevada of California is an instance. One adaptation which favours species which grow in fired areas is the ability to sprout quickly from underground food-storage organs unaffected by the fire: the European ling or heather (*Calluna vulgaris*) is a common example. Then there are species which are so well adapted that fire seems to be necessary for their reproduction: some of the species of pine of the Western Cordillera of North America have cones which only open and release their seeds when they have fallen to the ground and a fire has passed over them, heating them to a certain temperature, which is fatal to the seeds of some other species. Other advantageous features include the production of viable seed early in the plant's life-cycle (before there is another fire), hard-coated seeds and the ability, especially in trees and shrubs, to produce suckers from the base of the stem.

Since fire-resistant trees are relatively uncommon, repeated human-induced fire produces low vegetation types dominated by grasses and other perennial herbs rather than annuals or woody plants. Tropical woodland can at its arid margins be converted to savannah; woody savannah can become largely grassland. In the tundra, areas dry enough to burn may of course stay bare for a long time because of the slow rates of plant growth, but eventually lichens may replace species of moss. Animal communities are less affected unless their food supply changes, and this it may do for the better if there is more low foliage from shrubs which have replaced trees, or if grasslands contain more leafy species. On the other hand, the actual fire is anathema to most species and in fleeing they may fall prey to humans or to predator species: Australian aborigines cull many lizards from the edge of a bush fire just as African buzzards enjoy the feast of locusts and grasshoppers at such times.

There is little doubt that fire is of immense antiquity as a tool of environmental management by human groups. Natural sources of fire such as volcanoes and lightning are presumably the ultimate source, and exactly when the degree of control needed for successful landscape manipulation was acquired is still uncertain. It was certainly available to *Homo sapiens* from Late Pleistocene times onwards. In most modern western societies, the attitude towards it is largely negative because of damage to property and persons: thus immediate fire-fighting is often the response in suburban and wilderness areas alike, whereas controlled fire management might be more beneficial. Suppression at the very least tends to provide a reserve of fuel so that when accidental or natural fire breaks out, its effects are much more severe than if smaller controlled fires had been a feature of the land management systems.

Simplification is another set of alterations associated with human communities. It can result in sub-natural, semi-natural and cultural ecosystems. The process starts with human behaviour which for instance stresses one or more species to the point where their reproductive capacity cannot keep up with the rate at which they are being consumed by the human group. Alternatively, or indeed simultaneously, a species may not be extirpated but finds its competitive abilities weakened by the human-induced pressures so that another species takes over its niche in the system. The end result in each case is an ecosystem with fewer species than in its natural state, and natural processes will usually tend to revert to a more 'natural' state if the human stresses are relieved. Thus if the simplified state is highly valued by the human groups, they may have to put some effort into keeping the system simple, just as with the diverted state described above.

An example of simplification is the extermination of mammal carnivore species. These have been seen throughout history as threats to many human groups, either in a direct sense or more often because of the toll exerted on domestic grazing animals or on herds of wild herbivores such as antelope or caribou. The conflicts thus engendered have been symbolized by the need in many societies for the young males to kill a predator to establish their manhood, a rite which in northern Europe today seems to have been superseded by either football or drink ('I could murder a pint'), but which reached its apotheosis in the circuses of ancient Rome. Removal of most or all of the larger predators in many instances produces relatively little change in their prey herds; disease eventually brings about a similar result. But sometimes the removal of predator pressure may allow the prey population to grow abnormally, eat out their food supply (especially that of a period of scarcity like winter or a drought season) and induce starvation and a population 'crash'. (Neo-Malthusians often point to likely human analogues in the context of modern medicine and rates of population growth.) Thus killing a predator species may result in herbivores in winter eating all the young saplings in a forest as well as the bark of the mature trees, opening them to disease. In this way first a sub-natural and then a semi-natural (in this case open grassland or scrub) community is formed.

Fire may also reduce diversity, since not all species can tolerate its effects. There are two kinds of intense fire: crown fires and ground fires. The former are especially characteristic of coniferous forests and sclerophyll (Mediterranean-type and *Eucalyptus*) vegetation. Such a fire reduces the species of woody vegetation for some time, though eventually regeneration is likely and may in fact increase the number of herbaceous species. Ground fires are, over centuries, more manipulative of species and the temperatures reached ensure that only species with certain adaptations – as described above – will survive repeated burnings. So, for example, since the middle of the nineteenth century the drier moors of eastern England and Scotland have been converted into a semi-natural monoculture of heather (*Calluna vulgaris*) by burning in strips every eight to fifteen years. This has meant that many other species of grass and low shrub have been eliminated, since the heather regenerates from underground roots which survive the fires. The purpose of this manipulation is to increase the density of Red Grouse for sport, a bird whose major food when adult is heather. The diversity of the moor is further reduced by predator control, which means shooting rats (which are fond of grouse eggs) and any bird which might take adult or young grouse or the eggs. Diversity of moorland use is also reduced by the enthusiastic removal of recreationists during the spring and after the 'Glorious Twelfth'. By these means the Duke of Devonshire's 14,000 acres of moor in Yorkshire averaged a bag of 3600 brace by 1900, whereas fifty years earlier 200 brace had been the norm.

Domesticated grazing animals form a kind of low-temperature analogue to fire since they too exert selective stresses on a pre-existing vegetation. A heavy density of non-discriminatory animals such as goats will of course remove most of the vegetation from an area, but this is a rare occurrence and there will always be one or more species that is unpalatable even to them by virtue of a toxic compound in its leaves or twigs, or an armour of thorns. In a heavily grazed area, such a species will soon form the total vegetation. Where there is a choice of plants to eat, different species may produce differential results. Cattle eat by pulling at leafy material, but in the case of grasses this often leaves the site of leaf growth, the meristem tissues, intact. The plant will thus grow up again but may be dependent upon an underground store of nutrients in a bulb or tuber. If grazing is repetitive then the plant may exhaust this store and die or fail to set seed. By contrast, sheep act more like lawnmowers, cutting off everything near the ground and hence exerting a much more manipulative effect than cattle: a look at a British upland with a heavy sheep density will reveal only a few species of grass and prominent among them the wiry *Nardus stricta*, which offers resistance even to the incisor teeth of sheep, as well as being distinctly unjuicy. Domesticates can therefore change woodlands to scrub and thence to open grassland, can convert leafy scrub to a community of thorn bushes, and can do this over centuries so that the endpoints of the processes are only apparent when irreversibility seems to have set in.

Agriculture is also a potent source of simplification. Its course is generally obvious: once a crop plant or plants has been selected then most other biota are identified as weeds or pests. An ideal crop in modern 'western' agriculture consists of regular rows of a gentically uniform crop (hence ripening simultaneously and reacting alike to processing), with bare soil in between the rows (that is, no weed species to compete for nutrients or to harbour pests), no pests nor predators, and no surrounding vegetation such as hedges or copses which might harbour the latter. To produce this, a great deal of energy has to be expended and currently most of it comes from fossil fuels: no great wonder that some potatoes taste of diesel oil. In an ecological sense, therefore, a cultivated field resembles an early stage of a succession and (see table 2.1) is unstable; hence the heroic efforts of farm managers and producers of agricultural chemicals.

Simplification can also be extended to aquatic environments. Since fish have the convenient habit of swimming in shoals, it is possible to use modern technology to cull more individuals than can be replaced by normal reproductive rates. Although especially a feature of the steam winch and the nylon net, this has been possible for centuries in enclosed waters: many Pacific islanders have diverse conservation practices which are a clear response to over-fishing. Alleviation of stress does not always allow recovery, for there are instances of another (and usually less desirable) species taking over the station in the food web and permanently replacing the over-used species: heavily exploited sardine species off the coast of California have undergone this shift.

Obliteration extends both the above two processes; in fact it takes them both to what can be perceived as an ultimate stage, in which diversity is entirely removed by diversion. Fire, grazing and agriculture are instances of procedures which may all produce more bare ground than vegetated area. This is commonly associated with semi-arid areas where it is a component of the process labelled desertification. Its local ecology results from the harvesting of trees for woodfuel and the depasturing of domestic beasts at densities which cannot be sustained by the vegetation, a series of events often intensified by the boring of tube wells by development agencies which concentrates the animals spatially. Soils without a dense network of plant roots are especially susceptible to windblow in dry places. In such areas, when there is rain it often falls very intensively and so gulleys form and may develop into branching systems which once launched are difficult to check. In the wetter lands of upland Britain, our example of grouse moor fires shows that soil loss is not confined to arid zones. Burned patches often experience sheet erosion, and accidental burning of such areas may wreak enormous havoc over large areas, with fires smouldering on for months in the peaty subsoil and hence creating temperatures which inhibit plant regeneration but do not prevent further mineral soil loss.

However, any of these lifeless areas may only be temporarily so, since the surrounding areas may furnish seeds, bacteria and fern spores, and indeed any remaining soil may have a viable seed bank in it. Thus the notion that 'nature abhors a vacuum' may provide a sense of the onward impetus of succession. Nevertheless, erosion of the mineral soil may be a widespread and apparently irreversible process which leads to virtually lifeless areas. Sheetwash, gulleys and windblow alike make it very difficult for plant life to get established in a form which will halt the process of soil removal: algae and bacteria, for example, are usually powerless to do so. Erosion from area A may of course mean deposition at area B, which is consequently enriched; for example, many Mediterranean valleys are said to owe their fertility to the denudation of the surrounding hills. The many forms of terracing in the world are mostly attempts to come to terms with soil loss off slopes in agricultural economies. However, too much silt in the rivers causes them to flood and at the coasts sand-spits are formed, enclosing lagoons which in the past have often been malarial. Coping with soil loss is a complex affair which is generally much easier in a rich and technology-oriented society, where contour ploughing, no-tillage cropping, set-aside, rotation, windbreaks and use of exotics like the fast-growing Kudu vine all seem rational. Under the everyday pressures of Third World areas few of these techniques have much chance of, so to speak, taking root.

Industrial processes have generally produced areas which were inimical to life, at least temporarily. We can imagine the slag heap from a medieval bloomery only slowly being vegetated, or the lifeless water downstream from a lead smelt mill or a tanning works. Even the organic richness of a town midden may carry only limited and often undesired forms of life. After the nineteenth century, though, industry became bigger and faster, so that the production of more metals and fuels, rock and gravel, and their refining has meant the output of many substances which are toxic to most life forms. The refining of metals is perhaps the most obvious example: downwind from the stacks of nickel, copper and chromium refineries, for example, there is a plume of virtually lifeless land; if the plant is abandoned then it is many years before any visible vegetation can succeed, and indeed actual genetic mutation of plants has often to occur to produce a subspecies which is tolerant of the high metal levels. Loose material is analogous to sand dunes, where only a very few species can tolerate being constantly buried. But eventually some plant species will decide that it has found a post-volcanic bare zone or a glacial outwash plain and start to flourish. Once again, this process can be technologically assisted in the right circumstances: there are not many bare mine tips in Co. Durham now, but very old tin waste on the Jos Plateau in Nigeria is still untreated and virtually unvegetated. Contrariwise, chemicals unknown in nature (the products entirely of synthesis by technological means) rarely have breakdown pathways which would relieve their toxicity, but they tend not to be left around in large quantities in the environment; not legally, anyway.

The city is an obvious candidate for a figurative Waste Land, but in fact most urban areas have their own distinctive biogeography with few lifeless areas unless industrially inquinated. The output per unit area of fruit and vegetables from a cropped area can nearly always be increased by building suburban houses upon it.

Domestication is the process of genetic tailoring of species by humans. This means changing the genes so that any desirable characteristics are passed on to later generations of the plant or animal species concerned. Breeding has stressed various elements in the genetic potential of living organisms, from such obvious utilitarian features as increased yield in cereals to the rather less practical breeding of very small but extremely hairy pet dogs. Overall, higher yield (emotional if not nutritional) for less human effort is the aim of breeding programmes. Domestication has been a continuous process from some indeterminate period in the Late Pleistocene or early Holocene to the present day, when all the resources of science and technology can be deployed in plant and animal breeding, genetic engineering and genetic conservation.

Domesticated plants are the great global foundation of human nutrition and hence are the immediate reason why so much natural vegetation has been converted into cultural systems, either directly or via some phase of transformation of sub- or semi-natural areas, often called 'colonization' or 'reclamation' irrespective of the value of such areas for other purposes. Measures to increase output of roots, tubers, fruits, leaves and nuts are employed, but above all improved yields are sought of the seeds of the domesticated grasses we call after the goddess of agriculture herself, the cereals. The beginning of this process is not at present accessible to environmental archaeology, but it may have started with hunter-gatherers engaging in husbandry of patches of wild grasses that they came across in the course of food-gathering: scaring off the birds for a while, perhaps, or diverting a stream to provide more water. The key change comes when selected grains are used as seed-corn for a particular effect, such as size of seed-head or perhaps synchronicity of ripening so that all the ripe heads could be gathered on the one visit. Breeding for a brittle stem seems to have facilitated the use of tools like the flint sickle for harvesting and, presumably as a side-effect, the domesticated strains became annuals rather than retaining the perennial condition of their ancestors.

Breeding has tried to introduce many different characteristics at different times: resistance to pests, toleration of saline soils and droughts were all emphasized even before the nineteenth century and enabled many grassland and forest areas to be colonized for their production. Large areas of river valley and wooded slope were also converted to irrigated land. Few transformed areas were strictly natural when they were ploughed up or turned into *padi*, but they may have been quite remote and only manipulated by shifting cultivation or grazing. But other economies may have been displaced and so, indirectly at least, plant breeding was responsible for encroachment upon wild areas. All agriculture requires a lot of input energy in preparing the ground by sowing, weeding, fencing, harvest, food storage and preparation, and in the twentieth century this has in developed countries been largely that of fossil fuels, though in the Third World that of the sun channelled through human and animal muscle is still dominant. Given the need to restrain carbon dioxide output into the atmosphere we may all need to revert to that stage before too long.

Animals may, according to some archaeozoologists, have been domesticated before plants. The details are not as significant here as the importance of domestic animals as users of the margins. In a literal sense they can graze the margins of fields and woodlands where the plough cannot go or the irrigation channel cannot reach; more figuratively, they can find nutrition on the arid margins of the great river basins of the world, above the limits of potato culture on the Puna of the Andes or replacing the cattle and peasants of the Scottish Highlands with sheep and English lairds. Most domestic mammals, like their wild ancestors, can turn plant material which is either inedible to humans or so low in energy content as to be not worth the effort of gathering for direct consumption into a product useful to humans – themselves. With domesticated animals we have beasts which can be corraled or herded rather than hunted and whose reproductive habits can be carefully controlled, for instance by castrating those males whose blood lines are not considered helpful and thus producing fat and docile individuals. The animals function as moving providers of traction power, personal transport, leather, fertilizer, fuel and food. The latter comes as flesh and blood, but above all as milk, which is a renewable resource and moreover one which is amenable to storage once fermented or partially dried. Their ambulatory qualities confirm their utility in using marginal habitats and once again the wilderness becomes accessible and potentially productive.

We are about to experience an explosion of genetic engineering. The consequences for the wild might be twofold: first, that areas of natural (and thus inexpensive) storage of genetic variety in an evolving state may become ever more valuable as a pool of genetic material; second, that they could be one of the sites into which engineered organisms are released accidentally and find it possible to become naturalized.

Diversification is one of the aims of genetic engineering, but the process of increasing the number of species present in any one place has a very long history. Humans have been responsible for the transport of organisms around the world for as long as they themselves have moved: diseases are the obvious example. So there are many species of all kinds which have been accidentally introduced into parts of the world where they were not 'naturally' present, as well as those deliberately transported to alien climes for a number of reasons. Sometimes an introduced species needs a lot of effort to maintain it in its new home (think of orchids brought to Europe from the tropics), whereas others become easily naturalized so that self-sustaining populations develop, as with some species of Himalayan rhododendron in Europe and North America. Some get on so well in their new environment that they are classified as pests.

The methods by which accidental introductions take place are very varied. If an animal is introduced, then often its internal and external parasites and diseases come along as well and may make a transition to some other species (which is unlikely to have any immunity) in the new place. Human illnesses are very good examples: isolated populations of hunter-gatherers often seemed to have been free of influenza and measles, for instance, before contact with Europeans, and their populations were badly affected by the infections when they arrived with traders and missionaries. Larger organisms may appear in cargoes or in ship's ballast, for example, and most do not persist, though a few weedy species brought in to Britain in wool cargoes from Australia have become established in the flora. The tendency of rats to leave even viable ships is well known, and the European species (and their ticks, fleas and endoparasites) jumped to many an isolated island during the great explorations of the sixteenth and seventeenth centuries.

Deliberate introductions also have a long history, though for plants it was much intensified in the nineteenth century when means of keeping plants alive and salt-free on sea voyages were devised. A particular function was usually envisaged for each species, which might have been decoration or prestige as well as profit. The history of introductions goes back as far as there is evidence: sheep must have been introduced to western and northern Europe during prehistory, since there is no trace of a wild ancestor in those regions. It is usually thought that the rabbit was brought into Britain from Normandy as an easily managed source of meat, and its history in Australia is too well known to need repetition here. The house sparrow was taken to North America, apparently, as an act of nostalgia, as was the starling. The fauna of several islands in the Pacific and Caribbean was initially diversified in the sixteenth and seventeenth centuries by leaving goats and pigs on them to provide fresh meat on a return journey. In each national fauna and flora, too, there are species which it is suspected are the work of human introducers but which are so old that no firm date can be put to them and they have every appearance of being part of the natural biota: the sycamore tree in the UK is one such example. Of the common pheasant, all that can be said is that it is an introduced species (being a native of Asia Minor) brought in sometime before the thirteenth century.

The diversification produced by introductions may prove to

be temporary, since the new species may before long begin to oust other plants and animals. The rats on isolated islands, for instance, wreaked havoc on the populations of both flightless and ground-nesting birds; goats have often extirpated leafy and edible species from particular islands or island groups. But the opposite may occur and the new species may live alongside the old in apparent ecological harmony, like the wallaby population of north Staffordshire in England or the crested porcupines of Devon, both escapees from zoological gardens or wildlife parks. The point for our theme is that even remote areas may have introduced species which have every appearance of being part of the natural order, as indeed they probably are in terms of today's ecology. Yet they are biogeographically displaced and most directors of managed 'natural' areas try to keep them down and to prevent any more incursions from the more humanized parts of the world.

Conservation thus takes up the story. Basically, this is the attempt to exclude from a few areas of the globe a selection of human-induced processes, usually those of 'development', which can be seen as pushing a given land use further along the spectrum from natural ecosystems and wilderness to sub- and semi-natural, cultural and built environment. The process of conservation does not have to start at the 'natural' end, for the ideas can be applied to sub- and semi-natural areas as well and indeed to cultural landscapes, as when, say, exhausted gravel pits are turned into nature reserves. This general purpose for land use is commonly known as *in situ* conservation and contrasts with *ex situ* conservation, which is the protection of species or at least their genetic material in zoos, gardens, gene, seed, sperm and embryo banks. This latter has relevance here only in the sense that it may be a source of plants (but mostly of animals) for re-introduction to the wild when the ecology of an area newly under protection is being reconstituted.

Conservation as a process involves two main phases: (a) legislation and (b) management. The first is basically directed at *keeping out*, that is, excluding economic or other processes (like heavy recreational use) which would be detrimental to wild ecosystems, whatever their status in terms of 'naturalness'; the second is aimed at *keeping in*, that is, at perpetuating the ecosystems and their components into an indefinite future. Once the area is established or the species placed under legislative protection, then management is likely to be required. This may, however, be 100 per cent 'hands-off' management because the area is so remote that it can be safely left to itself (there are not many of these now, but perhaps parts of the Yukon might qualify as an example), or because its size effectively buffers some core zone of interest from all other influences, or simply because there is insufficient money to pay for the managers, the necessary equipment and the transport.

Where there is deliberate management of ecosystems, habitats and populations, then many factors are interlinked. First, there must be a long-term purpose for the conservation area, in terms of exactly what it is desired to perpetuate. This may sometimes be, for example, a particular stage of a succession because a valued plant or animal is characteristic of that particular stage. In this case, habitat manipulation may have to be severe in order to keep that stage in the reserve. An example would be a pond which housed a rare species of duck during migration periods. The pond, as ponds do, undergoes succession which covers the open water with plant leaves and provides less and less space for the duck to land on and to feed in. Management aimed at providing for the duck will then employ the local unit of Army Engineers to detonate explosions in the pond during the off-season to blow out the plants and set back succession by twenty years or so (I am not inventing this: I have seen it happen). So size, history, succession phase, resilience and human access are all factors which need scrutiny: even Antarctica has needed stringent management under the terms of the Antarctic Treaty of 1959 and its Agreed Measures.

Management of conservation areas is often directed at the control of human access for recreation. Anything which minimizes disturbance and impact is usually favoured: this may include maps which encourage people to go one way and not another (the use of colour can be persuasive here), fencing and signs, the use of boats and horses as rapid transit systems, or even bringing people in and out by helicopters or light aircraft. The more remote the area, the greater the problems access presents to the manager. If litter is left then it is expensive to get it out, if people are ill they are also expensive to get out (and had better be insured), and there is often the question of fire control. Fire is set by lightning in some natural ecosystems and suppression of it may deflect the ecosystem's normal development. In coniferous forests a fire suppression policy will allow perhaps one metre's depth of humic material to build up on the forest floor, and when there is an uncontrolled fire the mortality of trees will be much greater, since the temperatures will be very high and crown rather than ground fires will result. Yet the alternative – to allow conditions in which a party of Boy Scouts may be consumed by a fire, however natural – is not generally seen as creating a favourable image for conservation.

The most complete type of conservation area is the UNESCO-designated Biosphere Reserve, which is generally a zoned area with some parts open for access and recreation but where other zones are for the biota only, although some limited access for scientific research may be permitted. The whole is surrounded by a protective buffer zone in which agriculture, for instance, may be permitted if it is of a low-intensity 'traditional' kind. The High Tatra Mountains of Slovakia are one European example. One aim is eventually to have a network of such areas in all the world's major biogeographical provinces.

Being natural: apparent or real?

Given this list of factors in which human communities, often without any complex technology, can manipulate ecosystems, it is instructive to look at some wild places in the world which are associated in popular perception with the wilderness idea. We will examine two types of extratropical habitat to see if their pedigree as natural ecosystems can be proved, or whether they should be placed in one of the other categories.

The *Boreal coniferous forests* are a good example. They are distributed across the northern lands of North America and Eurasia (figure 2.1), south of the treeless tundra but abutting on the south with deciduous forests and with grasslands. There is no equivalent in the southern hemisphere. Even if these forests were cleared, the soils and climate mean that agriculture would be impossible, nor are domesticated animals grazed there. Among other factors, permafrost is a feature of most of these soils, as well as upper

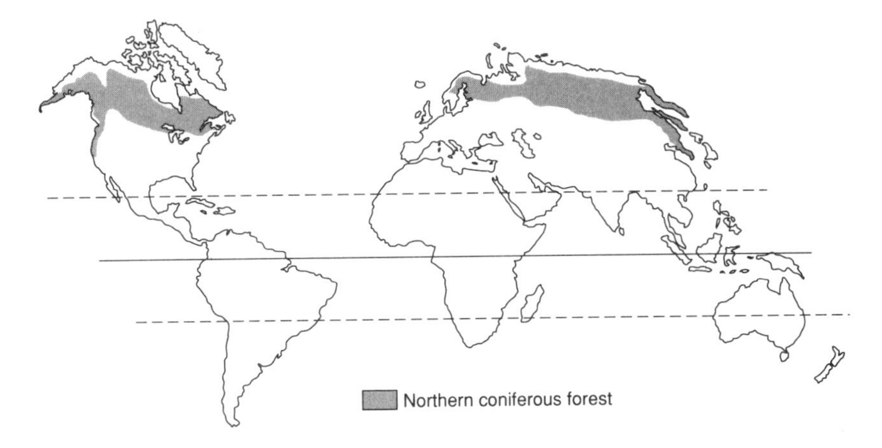

Figure 2.1 The distribution of the Boreal Forest biome.

forests are great swathes of wetland terrain of many ecological types, of which the Canadian muskeg is the best known. Large areas of forest are used for the timber and pulp industries, and post-1945 human incursions into the forest for energy development and defence installations are obvious. The rest is inhabited by the remants of aboriginal populations, living at low densities. Apart from the areas recently altered or utilized by twentieth-century technologies, then, is the huge stretch (some 12 million square kilometres) of the Boreal Forest a true wilderness, of the kind evoked so powerfully by, say, Sibelius' Fourth Symphony?

The biogeography of the forests is dominated, as is the ecology, by four genera of trees: pines, spruces, firs and larches. These are all needle-leaf conifers, but broad-leaved trees are also found, usually as successional species: alders, birches and aspens predominate. The decomposition of the litter layer is confined to a short summer period and so a thick layer of needles and twigs is common. Together with dead needles remaining on the tree for some time, this makes for a high susceptibility to fire of all three kinds. After fire, the usual succession (though varied by such factors as topography, soils, drainage, time and intensity of burn, and the condition of the surrounding woodland) is from bare ground–lichens–herbs–birch or aspen–spruce or fir. If the fire has left an undamaged layer of humus, seeds and roots, then both

Figure 2.2 Part of the actively managed Boreal Forest biome in Dalarna, central Sweden. This forest is actively managed (see the stumps of the harvested trees), and the dry floor, together with such management, prevents the build-up of litter to the point where the fire risk is exacerbated. There is scarcely enough fuel here to initiate the sort of conflagration that leads to the burning of the tree crowns.

(*Photograph*: I. G. Simmons.)

the speed and course of the succession may be different from that model. The bogs generally start as open water and have in postglacial times pursued a course towards dry land, though the infilling process (mostly with organic remains) is slow.

Within this forest tree and bog framework, two groups of animals are most visible to human perception. The first is the arthropods, with at least fifty common species including many aphids, the egregious blackflies and the budworms which present a hazard to commercial exploitation. The second is the mammals such as caribou, moose and beaver, along with predators such as mink, wolf and marten. The beaver and other fur-bearers probably played a greater role in the European-related politics of the North than any other animals anywhere until the twentieth century. By contrast, the fauna of the litter layer is less conspicuous and has to be very specialized. It includes, for example in Quebec, the aptly named millipede *Underwoodia* and the snail so intriguingly named *Discus cronkhitei-catskillensis*. Human occupation of the forests is not in general well documented archaeologically, but the youth of the ecosystems in postglacial terms presumably means that they have held human populations (with hunter-gatherer economies) for all the time of their existence. The human population densities on the Arctic slope of Canada before European contact were of the order of 0.1–0.2 per 10 square kilometres, and in the Great Lakes–St Lawrence drainage basins, 0.5–1.2 per square kilometre.

Given the ecology of these forests, it is not surprising that detailed palaeoecological investigation reveals evidence of a long history of episodes of fire.[2] Lake-bottom sediments are layered and the recovery of pollen, charcoal, mineral inwash and seeds, allied to radiocarbon dating, can give a good picture of forest development and recession. A study in Wisconsin, for example, showed that during the period 2000–1150 years ago, fires occurred within the catchment of a lake basin every 100 years on average; between 1150 and 120 years ago, the average was 140 years. A similar study in south-west Nova Scotia gave frequencies of 250 years in the period 6600–4550 BP and 350 years in the period 4450–2100 BP (Before Present: where Present = AD 1950

[2] R. W. Wein and D. A. Maclean (eds), *The Role of Fire in Northern Polar Ecosystems* (Wiley for SCOPE, Chichester, 1983).

in the conventions of radiocarbon dating[3]). One of the key differences in these different fire frequency cycles is the speed of postglacial immigration of deciduous trees from the south, since they are less susceptible to fire than the conifers. Research has also suggested that long-term climatic change is also relevant, with drier and warmer periods permitting greater fire frequencies.

Yet, as we know, these North American forests were not devoid of human presence in the pre-European era. Evidence from anthropological work shows how human communities such as the Beaver Indians of northern Alberta had a sophisticated and delicately tuned approach to creating fire. Certain patches of vegetation were burned deliberately in order to maximize their value as resources. Openings or clearings ('yards') were created within a forest area and maintained by burning; grass fringes of streams, wetlands, trails and ridges ('corridors') were similarly created and maintained, for they were both areas where hunted species of animal would either collect or traverse, or both. Fires were also set along traplines, around lakes and ponds, and within large areas of dead fallen trees which otherwise had no resource value; indeed, they were a danger since if ignited in summer they might start off a crown fire, whereas the Indian groups controlled time and place so as to produce only surface fires. So the yards and corridors may well have existed alongside a natural fire-produced mosaic, or could have used natural patterning as a starting point and maintained a version of it.[4] Presumably some of the fires registered in lake sediments are human-initiated, though it is possible that they have been too small a set of events to be visible in pollen and charcoal sequences.[5] Informants told one anthropologist that

[3] A. M. Swain, 'A history of fire and vegetation in Northeastern Minnesota as recorded in lake sediments', *Quaternary Research*, 3 (1973), pp. 383–96; D. G. Green, 'Fire and stability in the postglacial forests of southwest Nova Scotia', *Journal of Biogeography*, 9 (1982), pp. 29–40.

[4] H. T. Lewis, 'Maskuta: the ecology of Indian fire in Northern Alberta', *Western Canadian Journal of Anthropology*, 1 (1977), pp. 15–52; H. T. Lewis and T. A. Ferguson, 'Yards, corridors and mosaics: how to burn a Boreal forest', *Human Ecology*, 16 (1988), pp. 57–78.

[5] G. P. Nicholas, 'Ecological leveling: the archaeology and environmental dynamics of early postglacial land use', in G. P. Nicholas (ed.), *Holocene Human Ecology in Northeastern North America* (Plenum Press, New York and London, 1988), pp. 257–96.

their fire cycle was of the order of thirty-five to fifty-five years.

Thus only the extensive wetlands might merit the label of natural ecosystems; the forests are likely to have been subject to manipulation by human societies bent on creating a greater abundance of plant and animal resources and making possible a higher degree of proficiency in hunting.

The *moorlands of England and Wales* are another interesting example. They are often referred to popularly as the 'last great wildernesses' of those countries, and are perceived as being in a natural as well as wild state. Indeed, some are thus enshrined in the National Parks and Access to the Countryside Act 1949 as being areas of 'natural beauty'. Large areas of moorland terrain over 300 metres ASL exist some distance from roads, as on Dartmoor and in the North Pennines, and so the landscape is near as we are likely to get to wilderness in as small and densely populated a place as the British Isles.

Figure 2.3 A heather moor in upland England. There are a few trees by a streamside, but otherwise the scene is dominated by low bushy heather (*Calluna vulgaris*), whose density is produced by active management from grazing and fire. (*Photograph*: I. G. Simmons.)

That the moors have been invaded by the apparatus of the twentieth century is immediately apparent in many places. Quarries, military training grounds, recent afforestation schemes, enclosure for agriculture, microwave relay towers and even a BMEWS radar station have all been found on these moors. Yet enough wild area persists for the appellation 'natural' to spring to many minds. The vegetation is dominated by wild species of grass, sedge and low shrubs (like heather, *Calluna vulgaris*), with an occasional windswept shrub of rowan or thorn; large flat areas have a thick skin of peat ('blanket bog') which may be dominated by mosses such as *Sphagnum* species, so that in total it looks as if the upland climate is in control. If, however, we apply the same paleoecological analyses as in the North American forests discussed above, we find a history which tells a story of the interaction of human and natural factors in producing a landscape which is far from natural. Indeed, semi-natural is the applicable label in our terms. The tale starts with a bare tundra-like terrain at the end of the Pleistocene and proceeds through the early postglacial phase with a succession of woodland stages until almost all the uplands of England and Wales are covered with a mixed deciduous forest in which oak is the dominant tree and hazel a predominant understorey component. Other tree species were present, though in smaller quantities. Though it is difficult to map tree-lines from pollen analytical data, it seems very likely that only the highest summits of most uplands, and perhaps some exposed west-facing slopes, escaped the high tide of deciduous forest which culminated in about 8000 BP. Thereafter, it seems as if many human communities used the forest either as a resource *per se* or as a reservoir of land convertible to agriculture, and that the combined effect of these processes over many millennia was to produce today's vegetation; climate and climatic change has also been involved in the direction taken by some of these transformations.

It is not clear how far an oak woodland can be burned by lightning in the manner of the conifer forests described above. If the mid-postglacial climate (say 8500–5500 BP) was warmer and drier than at present, with more convectional rainfall, then the combination of more thunderstorms and a drier woodland may have allowed a mosaic of forest and successional stages. To this must always be added open areas on landslips, river banks and where trees have died or been blown down. What, however, seems to be clear from pollen and stratigraphic analyses of peats and lake deposits is that the terminal hunter-gatherer cultures (usually called Mesolithic in the British Isles) took a hand in managing the vegetation. Presumably they had in mind the same end as the Beaver Indians, though with (we assume: there is no direct archaeozoological evidence) red deer mostly in mind, and possibly with wild cattle as well. Evidence exists for small clearings associated with fire whose siting and frequency make lightning an unlikely cause. Some of these clearings colonize up to forest again, but in this cool and wet climate others suffer from waterlogged soils once the 'water-pump' effect of the trees has been removed. Hence we see the beginning of a blanket of peat on parts of the uplands, which came in time to be as much as four metres deep.

Once agriculture had begun in Britain (after *c.*5500 BP), the forests changed their role. Burning for animal fodder seems to have ceased at the end of the Mesolithic, and this in places allowed forest to recolonize shallow peats. But then much more forest disappeared as clearings were made for agriculture. Some cultivation was of the shifting type, but yet more as time progressed involved the keeping of permanent and fertilized fields. A climatic deterioration during the Iron Age may have assisted in the downward movement of settlement, and progressive soil change must have promoted the same shifts, so that by medieval times we see the upland used little for agriculture and more for pastoralism, with only a fringe of woodland left on steep slopes.[6] Much of this woodland was itself subject to intensive management in areas where for example ironstones were discovered and the iron-producing bloomeries were fuelled by charcoal.

Pastoralism involved the keeping of sheep and cattle on the upland pastures. The relative numbers of these two domesticates has altered the composition of the vegetation since they have different grazing habits. Where, as now, sheep are dominant then mostly leafy grasses are grazed out and wiry species take their

[6] Since this was the time of the great Cistercian monasteries, and since the history of the woodlands was to regress from small bare patches in the Mesolithic to a fringe round the steeper slopes in the medieval period, I have called this the 'tonsure model' of forest development.

place. The representation of bracken may also be in part affected by grazing, since cattle will eat the young fronds. Climate is also important here since bracken is only partially tolerant of frost and intolerant of low levels of soil oxygen.

A last ecological influence (previously discussed in outline) dates from the nineteenth century on the drier moors of the eastern side of the uplands, including Scotland. After the 1850s, a change took place in the shooting of red grouse. Instead of walking up the birds with dogs and firing at them as they flew away, estate managers took to improving the density of grouse by burning moors in strips at 15–20-year intervals. This produced a near-monoculture of heather, but the strips ensured that each pair of birds had within their territory some plants of different ages, for food and for nesting cover. Large numbers of grouse could then be put up by beaters and flushed over a line of butts where a sportsman with a double-barrelled shotgun and a loader might wreak slaughter of a gratifyingly quantitative kind. Ecologically, the result has been a moorland dominated by heather and the consequent landscape is sold as an attractive feature of wild scenery; but it is very largely a man-made (*sic*) habitat.

So if we look at the moorlands' history we can see that in no case can we grant the certificate of pristine ecology even to the areas not obviously affected by settlement, extractive industry or recent afforestation. This does not prevent them having a high valuation in a country where really wild areas are scarce, but it does mean that they do not qualify as natural ecosystems within any accepted definitions of those words.

A depth of change

Inspection of these examples of seemingly natural but in fact considerably altered ecosystems prompts us to look more widely afield in time and in space: have there been human effects on all the major world biomes, or can we still point to truly pristine ecosystems? If transformations have been brought about by human societies, are these artifacts of recent years or may they go back as far as the earliest periods identified in chapter 1? In chapter 3, we will look at the evidence for some of these metamorphoses.

3
‘*A culture is no better than its woods*’: The Humanization of the Wilderness

In this chapter we shall examine some non-urban areas of the world to see how and when they have moved away from a natural condition to one in which any areas of natural ecosystems are small and scattered, or even absent altogether – that is, they have become humanized. The examples are chosen to reflect different scales of time and space.

When were the first non-natural landscapes created?

The far past, and the origins of things, have a special fascination. So it is with the human alteration of landscapes. If we consider the genus *Homo* in the Pleistocene period, then to be able to transform the natural communities of plants and animals into near-natural or semi-natural states, hominid groups would have needed to have taken one of two paths. The first would have been the ability to deplete a local population of animals or plants so thoroughly as to affect the local ecosystem for archaeologically detectable periods of time. The consumption of a particular plant species beyond its growth and reproductive capacities would be one possibility, and the other the killing of too many gravid females of a species of deer. The second tool for producing a humanized landscape is fire, exerting the effects discussed in the previous chapter.

Fire has been much considered as a possible manipulative tool in the hands of early hominids, for example in association with *Homo erectus*, as well as with Palaeolithic populations of *Homo sapiens neanderthalensis* and eventually *H. sapiens sapiens*. The evidence for the earliest periods comes from baked clays, occasional charcoal fragments in open sites, and charred materials in caves. Thus cone-shaped masses of burnt clay are found in association with remains of *H. erectus* at Chesowanja in Kenya, with an age of 1.5 million years ago (My) and a spread of sites with reported evidence of fire in China, Hungary, Mediterranean France and western England in the period 300–500 thousand years ago (Ky). A problem of interpretation is that the 'fired' materials at many sites have been inferred from their colour, when staining by manganese might give the same effect. Equally, the clay masses at Chesowanja might have been produced by a natural fire consuming a tree stump. It is also reported that natural fires can occur inside caves. These revised explanations led S. R. James,[1] who examined accounts of 30 sites in the Lower and Middle Pleistocene of Europe, Africa and Asia, to suggest that there were no unequivocal hearths until the advent of the Neanderthals at the end of the Middle Pleistocene (*c.* 80 Ky). Thus the deduction of a neat pattern of (1) learning how to use fire in Africa and then (2) needing it to colonize cooler areas may have been over-ambitious, especially if the leap is made from control at the hearth to regulation in the landscape.

Inferential evidence such as that provided by pollen analysis of lake deposits is also open to variable interpretation. At two English sites (Hoxne and Mark's Tey, in East Anglia), deposits of the Holstein Interglacial (220–200 Ky) reveal an episode some 350 years long in which dry forest disappears rather abruptly and is replaced by grassland and by some birch and pine. Charcoal was found at the appropriate horizon and at Hoxne some Acheulian tools, but no evidence of hominid bones. This episode

[1] S. R. James, 'Hominid use of fire in the Lower and Middle Pleistocene', *Current Anthropology*, 30 (1989), pp. 1–26. For the Hominoidea, W. Schüle locates 'getting used to fire' back in Mio/Pliocene times ('Human evolution, animal behaviour and Quaternary extinctions: a paleo-ecology of hunting', *Homo*, 41 (1991), pp. 228–50).

of forest recession has been variously explained: in general, biological scientists attribute it to natural fire, climatic change or high densities of browsing animals, whereas archaeologists see early evidence for fire being used by *H. erectus* groups to produce a near-natural landscape.

Just as inferential are the surmises (even when based on ethnographic analogy) of how useful controlled fire would have been to prehistoric humans: heating, light, nocturnal protection, roasted and dried meat, detoxified plants, less inedible plant materials, animals burnt or blinded by smoke, hardened spear-points and digging sticks and even honey from fallen trees might all be reasons why fire would be a popular possession of hunter-gatherer people. Small wonder, perhaps, that so many men even now go stand with their backsides to a fireplace with a fire in it. Indeed, Clark and Harris[2] suggest that fire as much as any other factor may have been responsible for the formation of the family units of human societies.

At the end of the Pleistocene, some 200 genera of warm-blooded animals with adult weight of 50 kg or more ('megafauna': mostly mammals but some birds as well) became extinct. This phenomenon is usually referred to as 'Pleistocene overkill'. Centred on *c.*13,000 BP, it seems as if the extirpation of these animals took place at about the same time as the introduction of humans into North and South America (33 and 46 genera extinct respectively), whereas in lands already occupied the changes were less: only 7 genera in Africa. In Australia, the body count was 19 and one site shows about 7000 years of stable coexistence of humans and megafauna. By contrast, human occupation of New Zealand and Madagascar, for instance, was much later and there the megafaunal extinctions too were much further on in time. Not surprisingly, this interpretation of the far past has been contested. The natural stresses of climatic and habitat change at the end of the Pleistocene are often cited as more likely explanations, with the process gaining momentum as predators found their prey species reduced in number. The answer may well lie in the complexities of animal populations struggling to survive in the rapidly

[2] J. D. Clark and J. W. K. Harris, 'Fire and its roles in early hominid lifeways', *African Archaeological Review*, 3 (1985), pp. 3–27.

changing natural conditions of the Late Pleistocene and then being confronted by a new predator against which they had evolved no behavioural responses for defence. Fire is called in to support both camps, since when we come to the Upper Pleistocene and *H. sapiens* is established as the only hominid, then the possession and control of fire is not disputed. So if fire was implicated in 'overkill' then it becomes likely that humans were involved in its controlled use.

In general, it seems as if the evidence for human-produced near-natural and semi-natural landscapes is much firmer in the Upper Palaeolithic, with the advent of *H. sapiens*, but the probable usefulness of it in earlier periods might have been so great as to make it a reasonable supposition (but no more) that its first use is buried in a much greater antiquity. It would be in the controlled use of fire outside the hearth in the local landscape we might find the first recognizably humanized landscapes of the planet.

The European wildwood

If remote sensing satellites had existed in about 7000 BC, then the pictures sent back to Earth of the land cover of Europe would have shown a certain monotonous quality, for in the lowlands, and even to a considerable height on the mountains, there was a blanket of mixed deciduous forest. Admittedly, this gave way to coniferous admixtures towards the north and upwards on the mountains, but at low levels of resolution it would have appeared broken only by the threads of the great rivers and by the shiny reflections of large areas of wetlands too soggy to bear trees, as in the western part of the Netherlands. But nowadays a flight on a clear day from, say, Dublin to Frankfurt shows that much of this forest has gone and that much of the existing woodland is of coniferous species. The wildwood has gone and with it the greatest chance that lowland Europe might still have some true natural ecosystems left.

If our imaginary satellite had been capable of high-resolution pictures then the images would have shown that the forest was not totally unbroken. Here and there the canopy had gaps where an old tree had succumbed to age and fallen; bigger gaps indicated where windthrow had laid flat several trees (or even large areas after particularly bad storms); river valleys might have had braids of flood deposits free of trees, and unstable slopes anywhere were free of forest cover. This kind of mosaic is of course the result of the natural processes within a woodland ecosystem. But in 7000 BC, even before the coming of agriculture to most of central, western and northern Europe, the mosaic in many areas contained another element: the openings, referred to briefly in chapter 2, maintained by hunter-gatherer groups of humans of Mesolithic cultures. Evidence from pollen and charcoal analysis of a wide variety of deposits from Poland and Italy to Ireland shows that these groups were associated with the maintainence of small clearings. They may also have created them, although the evidence for this is more tenuous and it may well have been easier to preserve existing openings which had a totally natural origin.

The virtues of such practices are again a matter of inference, although we have the ethnographic parallel of north-eastern North America as a guide to possibilities. Forests subject to regular ground fires develop little underbrush and so they are easier to traverse, and more open places in which to spot game and to get a clear line of fire for arrow or spear. Openings near water might well attract game to graze as well as to drink, and hence provide a handy locus for larder replenishment. To add to this, recent ecological studies have shown that after fire, successional species of shrub (as in Europe the hazel, *Corylus*) have foliage which is not only within the browsing distance of mammals such as deer but is higher in leaf protein than unfired vegetation. Yet further, shrubs like hazel produce a crop of nutritious nuts; *Corylus* shells are found in many excavations of Mesolithic sites. So the advantages of a partially-controlled mosaic of vegetation are manifold.

The first cultivators in Europe may not, then, have come into a totally natural landscape, though the scale of alteration away from the natural was of much the same order as natural processes. For small-scale cultivation of cereals and the keeping of domestic animals such as cattle, the interspersion of small clearings, areas of secondary woodland and remaining swathes of high forest was presumably a welcoming environment. This would have been the case whether agriculture spread by folk-movements carrying the new technique with them or by adoption of cultivation by indigenous groups. The new technology of the Neolithic period, however, included the polished stone axe, which is far more effective at directly felling trees than anything in the Mesolithic toolbag. Thus larger areas might be cultivated and a denser human population sustained, whether as effect or as cause. In many areas of Europe, the form of agriculture adopted was that of shifting cultivation (now most familiar to most of us in its recent tropical forms) where a plot was cleared and the slash burned, with the ash providing an initial input of fertilizer. The plot was then cultivated until either soil fertility fell or the task of weeding became too onerous; it was then abandoned and the secondary succession allowed to colonize the open area with shrubs, shade-intolerant trees and finally the closed canopy species of the high forest. From the air such an environment would have looked largely forested, but close inspection would have revealed a lot more secondary woodland than under Mesolithic occupancy as well as more openings. More smoke, too, we imagine.

Any group which kept domesticated animals would have had the capacity to affect the ecology of forest without directly felling it. Cattle, for example, will eat succulent shrub material and even bark. With every item of the former consumed, there is one fewer seedling that will become a mature tree; with the latter there is always the chance that the bark (possibly combined with rubbing) will be eaten off all round the stem and so the tree, effectively ring-barked, will die. A herd of cattle, therefore, is the agent of subtle but chronologically far-reaching metamorphoses in the nature and extent of forests. Domestic pigs, likewise, will consume much of the seedfall of a forest, in competition with squirrels and voles for instance, and in years of thin mast and acorn fall, there may be no seeds at all left to sprout and grow. In the Swiss Neolithic, it seems as if leaf foddering for cattle was practised: ash, lime and elm trees were stripped of leafy branches, and it may well be that management of trees to produce such shoots (for example, by shredding or pollarding) dates back at least to about 3000 BC.[3]

[3] P. Rasmussen, 'Leaf-foddering of livestock in the Neolithic: archaeobotanical evidence from Weier, Switzerland', *Journal of Danish Archaeology*, 8 (1989), pp. 51–71.

The ecology of a woodland may well be altered by use, even if that utilization is sustainable. In Neolithic peat deposits (*c.*4000 BC) in the Somerset Levels of south-west England, well-constructed tracks of wood have been excavated. Their construction makes clear that small timber trees of oak were used, along with large underwood poles of ash, lime, elm, oak and alder, and small poles of hazel and holly. The quantities used suggest, along with pollen analytical work, that woodlands were deliberately managed so as to produce underwood, chiefly by the practice of coppicing; this would have the advantage of yielding a lot of leaf fodder as well. Some tracks were made just like wattle hurdles, which require a great quantity of rods between four and ten years old, produced by coppicing a hazel wood in which there was a little ash.

The later prehistory of European woodland consists largely of more of the same, intensified by increasing human populations. Selectivity was possible for some groups: in the English-Welsh borderlands, it seems that Bronze Age people went for areas vegetated by lime trees (*Tilia* sp.) as prime sites for fields: they presumably knew that this species is most likely to be found on well-drained soils with a high base status. But eventually the human population numbers became such that permanent fields became necessary, with their fertility kept up by manuring. A core of such fields near the settlement became the focus for an estate in which managed grasslands supported cattle and sheep and the woodland beyond was for grazing and timber yield. Between this and the next estate there was a fixed boundary which has in some parts of Europe formed the basis for later civil divisions such as the parish and the commune.

With the advent of metal, localities which engaged in the smelting of ores found another demand upon their tree stock. In Roman times, for example, it took 84 long tons of wood to make the necessary charcoal to produce one ton of iron; one ironworks in the Sussex (England) Weald is estimated to have produced 550 tons per year in the years AD 120–240, so that in total 7.5 million tons of wood were needed, consuming the standing crop from 300 square kilometres, about 9 per cent of the whole region of the Weald. But as Oliver Rackham remarks, there is no need to think of metal-smelting industries as enemies of woodland, for trees will grow again and the whole of these industries could probably have been supplied with their needs from rotation coppice: industrialists have a keen eye for their future supplies of fuel.[4] Industry was often, before the nineteenth century, an agent of woodland conservation. The same can be said of oak bark production, which was the staple of the tanning trade. At the height of this business in Britain in the early nineteenth century, leather production required the supply of about 500,000 tons of bark per year, a figure which Rackham suggests is double that of British merchant and naval shipyards combined. In Russia, the important potash industry needed three cubic metres of wood per kilo of potash produced, and western Russia also produced iron and salt, with the Kama saltworks getting wood from over 300 kilometres away by 1750. Add agricultural colonization and we have a European Russia which has 53 per cent of its land surface under forest in 1725, 45 per cent in 1796 and 35 per cent in 1914.[5]

The connection between forests and sea-power has long been recognized,[6] though perhaps the connection has been somewhat exaggerated by the fact that naval yards may have sought much of their timber from parks or hedgerows, since only there were the oaks large enough and crooked enough to contain the special shapes demanded for large wooden hulls. In the France of the eighteenth century, the navy sought out all the usable oaks and firs and felled them without thought for the condition of the remaining woodland; this land was usually left without shade trees and so colonized by birch, which was in turn appropriated for pasture and fuel by peasants and hence the forest was reduced in area.[7]

A ground bass to all this time and activity is the humdrum business of supplying a pre-industrial society with fuel and with timber for building and pole-sized wood for a myriad of other

[4] O. Rackham, *Ancient Woodland. Its history, vegetation and uses in England* (Edward Arnold, London, 1980).

[5] R. A. French, 'Russians and the forest', in J. H. Bater and R. A. French (eds), *Studies in Russian Historical Geography*, vol. 1 (Academic Press, London, 1983), pp. 23–44.

[6] R. G. Albion, *Forests and Sea Power: The timber problems of the Royal Navy 1652–1862* (Harvard University Press, Cambridge, MA, 1926).

[7] P. W. Bamford, *Forests and French Sea Power 1660–1789* (University of Toronto Press, Toronto, 1956).

purposes. Further, such production had to be sustainable and preferably near at hand. The result is a series of manipulations of tree and shrub species, of which coppicing is one and practices such as pollarding and shredding are others. The feeding of cattle on foliage was important, as might also be the short autumn season of feeding pigs on acorns and beechmast, the pannage. One result of many of these practices was not necessarily the disappearance of wood entirely but the production of a mixture of woods, heaths and grasslands, collectively called wood-pasture; in fact domesticated animals had to be kept out of forests if this was not to happen. On wood-pasture areas, the quantity of trees rose and fell with the intensity of grazing. So even in an area which was totally wooded, the ecosystem might well be near- or semi-natural. It might be in a state of succession from grazing land to forest, for example, or it might be managed woodland where even the species of the ground floras reflect the felling and regrowth regime: the distribution and density of bluebells, for example, is determined by the onset of coppicing, and the opening of woodland canopy increases their vigour spectacularly in the second year after coppicing. Bracken increases from the first summer and may become the dominant plant of the ground layer.

So on the verge of the industrial revolution in Europe, there can have been virtually no virgin woodland left. The coming of an industrial economy produced a number of changes: coppicing declined because the railways brought cheap coal to rural areas and were replaced either by cropland or by trees for timber, to be used for furniture for example, as with beech in southern England.[8] The steamship made it possible to import large quantities of timber except in times of war, when fast-growing exotic species of conifer were planted. In Britain this shift dates largely from the period of the First World War, but in Germany pine and spruce began to replace oak and beech in the eighteenth century. The eastern German lands provided seeds of larch and spruce, and attempts to seed pine dated from the last twenty years of the previous century.

The moral of this discussion is clear: in Europe there are probably no 'virgin' woodlands at all. Those pieces shown to visitors as

[8] O. Rackham, *The History of the Countryside* (J. M. Dent, London, 1976).

urwald should not be believed in until positive vetting has occurred and a certificate (in the form of a paper in an internationally refereed journal) awarded. There are woodlands that are wild and unmanaged, there are those which stand on land where there has been woodland since time immemorial, and there are attempts to reconstruct primeval woodland; but these are not the same.

The forests of the tropical lowlands

No stretch of apparently near-natural vegetation is more under scrutiny today than the broad-leaved forests of the equatorial lowlands of South America, Africa and Indo-Malaysia. Often referred to popularly as Tropical Rain Forests but more recently in more specialized literature as TMFs (Tropical Moist Forests), the conversion of these great swathes of forest to other uses has excited attention on a variety of grounds, not the least of which is the findings of the modellers of global climate that their conversion releases large amounts of carbon into the atmosphere and that any replacement vegetation does not sequester nearly as much carbon as the original forest. A surplus thus remains in the atmosphere to enhance the 'greenhouse effect'.

In the light of this global role, the tendency has been to see the TMFs as great natural governors in the mechanical sense, keeping the climatic engine at a steady speed without either sudden surges or losses of power. For quite a long period, too, this image was reinforced by palaeoecological assumptions that they had beed *in situ* at least since Tertiary times and that their great species diversity was due to their not having suffered much change during the Pleistocene, unlike their temperate zone counterparts. These views have now been revised in the light of better scientific information, and it seems as if the Pleistocene episodes of cooling in the temperate zones were mirrored by both cooling and aridity in the Tropics.[9] Thus during the early Holocene the TMFs

[9] J. R. Flenley, *The Equatorial Rain Forest. A geological history* (Butterworth, London and Boston, 1979); P. Colinvaux, 'Amazon diversity in light of the palaeoecological record', *Quaternary Science Reviews*, 6 (1987), pp. 93–114.

expanded from refuges in the most favourable climatic spots just like the forests of Eurasia or North America.

Somewhere in the period 9300–8500 BP, the TMFs were established in their modern locations, perhaps even as late as 5000 BP at the margins. Do we then assume that from then until the recent period of rapid conversion of these forests they were true wildernesses in the sense of having no detectable human presence? A caution about the quality of the evidence is needed here since it is only relatively recently that both archaeology and palaeoecology have been producing the sort of evidence to which we have become accustomed in temperate and arctic regions, so valid generalizations are hard to achieve. For example, long-period palaeoecological studies which show the influx of pollen and charcoal into lakes and mires are much less frequent in the lowland tropics. Where they exist, it is interesting to see that evidence of fire in the lowland tropical forests of Venezuela can be detected for periods as long ago as 6200 BP; yet human presence begins *c.*3750 BP. It seems possible that even these forests can burn, possibly during climatic oscillations towards greater aridity.

Once human presence in the forests was established, and agriculture began, then some impact on the ecology was inevitable.[10] The pattern most often established was that of shifting agriculture (sometimes called swidden'), not basically different from that of Neolithic Europe described above. In spite of the fast growth of trees and shrubs in the lowland tropics, the time from clearing to high forest is usually at least 100 years and there are instances of perhaps 300 years being taken. When regenerated, the forest is unlikely to have the same species composition as the area which had been cleared because the species diversity means that there is a greater pool of tree species able to provide the eventual canopy and supra-canopy individuals. In one site in Papua-New Guinea there was clearing for agriculture at 4300 BP, only 200 years after the forest had assembled. There was partial regeneration 300 years later, but high forest was seen only at 3000 BP and it had a higher proportion of secondary species than the original eco-

[10] R. P. Tucker and J. F. Richards (eds), *Global Deforestation and the Nineteenth-Century World Economy* (Duke University Policy Studies, Durham, NC, 1983).

system. This must have been a common pattern at the latitudinal and altitudinal margins of the TMFs. For the whole of the tropical forest regions, there was probably the same situation as described for the Hanunoo of the Phillipines, who named about 1000 varieties of plants as being uncultivated but probably engaged in some form of husbandry of most of them. Further alterations would have been made by groups which did not abandon agricultural clearings to natural succession but instead planted the forsaken area with shrub and tree species of their own choice. Thus the conclusion of one ecologist, that in southern Nigeria the whole of what was (in the 1950s) continuous forest had been inhabited and cultivated at some time or other, probably has a very wide application.

If we take 7000 BP as a rough date by which crop cultivation had been introduced into the lowland tropics in all three major TMF regions, then there has been a long history of human-induced change before any influence of European contact can be seen. When it came, it was usually in the form of new crops which permit more intensive food production and higher population densities, leading to higher rates of soil erosion. The introduction of the sweet potato into the Philippines (whence it spread very quickly to Papua-New Guinea) by the Spanish in the sixteenth century is one example; general contact with the European-style cash crop economy in the 1930s produced even more ecological impact, as evidenced by rates of erosion of soil into basins where it can now be measured and dated. In regions with later European contact, the impact might be even greater and more sudden. In Java it was during the first decades of the nineteenth century that immense deforestation took place under European control, aided by the 1000–1500 tons per year of imported iron and 200–300 tons of steel, 'worked up into the implements of husbandry and into the various instruments, engines, and utensils required . . . in the various districts', noted by Stamford Raffles in 1817. And as a kind of undercurrent to all these changes, we note that the indigenous inhabitants had carried on trade for at least 5000 years in some places. The exports of forest products were exchanged for other agricultural outputs, brassware and eventually guns. Such processes must have had a gradually cumulative effect upon forest ecology as well as being responsible for the expansion of agriculture.

So in strict terms the TMFs were not wilderness even before the 1930s: they were inhabited and economically used, though in a manner which tended to allow the spontaneous regeneration of something which was to most outsiders closely akin to the original woodland. We might be permitted the slightly heretical thought, however, that the great species diversity associated with the TMFs could be related to it being a mosiac derived from centuries of human occupation and use. Where secondary vegetation persisted, it was replete with low shrubs and many climbers and lianes, thus giving these forests their undeserved soubriquet 'jungle' with its negative connotations ('the law of . . .'), as well as the setting for hundreds of tales of Tarzan. But this perceived wilderness (many Europeans evaluated it as a kind of very damp desert) was in fact a home for many humans and to a great extent their garden as well. Confirmation in the stark measures of today's economics comes from estimates that the value (to commerce, of one hectare over one year) of sustainable forest products from near- and semi-natural ecosystems in Amazonia is US$6820. That of the products of reforestation with the single tree species *Gmelina* is US$3184, and that for the yield of cattle pasture, US$148. The meat from the latter does not, though, provide a sustainable income.

The woodlands of eastern North America

The European discovery and conquest of North America is intimately tied up with the idea of wilderness. The early settlers and subsequent generations saw the great inland tracts of forest as a wilderness to be eradicated in the cause of creating civilization. For many years the frontier was a forested zone, and the remaining forests of North America still form the battleground for both that early ideology and its mirror image of a fierce attachment to conservation. It is common knowledge that westward settlement destroyed a great deal of forest and left cropland, grassland and industry behind; but satellite images and aerial photographs show a great deal of woodland still east of the Mississippi. In his book *Megalopolis* (1961), Jean Gottman remarked that even the Boston-Washington DC corridor had extensive areas of tree cover. Our task here is to examine the provenance of the eastern woodlands.

At the time of European contact, forests were the major vegetation type from the Great Lakes to the Gulf Coast and inland well towards the Mississippi and Missouri rivers. These forests were mainly of deciduous trees on well-drained sites, but wherever conditions were more stressed the coniferous element increased and along with it the frequency of natural fire. Thus upwards on mountains, towards the north, and on the sandy soils and swampy areas near the low-lying coast, needle-leaved species were common and in some places dominant. This forest was, as in Europe, the result of colonization after the glacial episodes of the Pleistocene and was probably not yet in equilibrium with climatic change: natural adjustments were still occurring.

Just as in the Old World prehistoric peoples have come to be recognized as being responsible for metamorphoses of forest ecosystems, so the indigenous inhabitants of eastern North America have come under scrutiny as ecological agents. At the time of the European settlements of the sixteenth and seventeenth centuries, the eastern woodlands were occupied by a variety of groups of cultivators. For them, domesticated plants (among which maize was predominant) provided about half the total diet, with hunting and collecting in the forest yielding the rest. Recent work suggests that the population growth of the Indians in North America may have been 90–112 million at contact time, a density as great as that of western Europe, and so the number of cultivated plots cannot have been small. In order to grow corn, beans and squashes, the Indians used fire a great deal. Small bushes were uprooted and burned, larger trees debarked or felled with stone axes with the stumps left in the ground. Plots might be as big as 200 square metres and the crops cultivated in rows or on mounds. Wood was used for houses and village size varied from 50 to 1000 in population. A study of the Ontario Huron showed that a village of 1000 inhabitants would need 16,000 poles (*c.*10 centimetres in diameter by 3–10 metres long), 250 interior posts 25 centimetres by 3–10 metres, and 162,000 square metres of bark as a covering. Added to these would be the timber requirements for a palisade and for fuel. Secondary woodland would best provide these kinds of underwood sizes, with abandoned cropland an obvious source of such secondary vegetation. The total cleared area for one of the larger Huron settlements was of the order of

245 hectares, and this area might increase as demands for fuel wood, for example, had to be met by felling trees rather than collecting dead wood. Around the settlement there was also a zone of heavy collecting of wild plant material (fruits, nuts, maple syrup and berries), to the extent that their density even today suggests that some form of preferential husbandry was carried on. Furthermore, it was the practice to move the whole village from time to time, either on a regular seasonal basis or, more often, when resources were exhausted. Early writers often referred to the frequency of 'old fields', 'flats', 'openings' and 'meadows', which were characteristically vegetated with grasses.

In the remaining woodlands, the Indians were active managers. In particular, they used fire to maintain openings in the forest in a grassland vegetation and to attract big game; to extend such openings where possible and to encourage the growth of fresh grass and of the secondary vegetation which was a high yielder of berries, nuts and small game; to hunt game, herding it together for slaughter by encircling beasts with fire; and for miscellaneous purposes such as warfare against rival groups and driving off biting insects.

The regular use of fire had two major sets of results.[11] In the forest itself, a scrubby understorey is discouraged and the ground layer tends to be grassy; fire-tolerant species come in time to dominate the tree species composition. Then the forest gets thinner until first a park-like vegetation and then grassland is created. Early settlers in the north-east noted that some areas were very 'like our Parkes in England' and became 'very beautifull and commodious'. In particular, this process resulted in large grassy zones whenever the forest was reaching a climatic boundary, whether longitudinally or altitudinally. Thus in Appalachia, the mountains might be capped by a bare area (the 'balds') even though the natural upper limit of forest growth had not been reached. Towards the west, the increasing aridity meant that there were thousands of hectares without any trees at all, to which the name 'prairie' became attached.

[11] W. A. Patterson and K. E. Sassaman, 'Indian fires in the prehistory of New England', in G. P. Nicholas (ed.), *Holocene Human Ecology in Northeastern North America* (Plenum Press, New York and London, 1988), pp. 107–35.

It seems clear that in the eastern woodlands the Indians were, to use the words of Williams's major[12] study of the forest history of the USA, 'a potent if not crucial ecological factor in the distribution and composition of the forest'. He goes on to write that 'the idea of the forest as being in some pristine state of equilibrium with nature, awaiting the arrival of the transforming hand of the Europeans, has been all too readily accepted . . . as a benchmark against which to measure all subsequent change.' That subsequent change is an often-related tale, with varying emphases according to time and place of telling. The story after the establishment and westward expansion of the Europeans has a number of general aspects which are worth remembering as evaluative of the ecological status of any large areas of forest in the eastern half of North America.

The first of these is the great inroad on timber made even by early agriculturally based settlers as they cut fuel for fires, for building, for potash fertilizer, for ship timber, naval stores and local iron-smelting. A household in the late eighteenth century might consume 4.5 cords per capita for fuel each year (a cord is a split and cut stack of wood $4 \times 4 \times 8 = 128$ cubic feet or 3.6 cubic metres), and so the amount of wood cut for fuel each year probably exceeded that used for any other purpose until about 1870, when lumber overtook it.

The second point to remember is that the industrial revolution in North America between 1810 and 1860 was largely powered by wood and water. Wood was used for the firing of boilers on steamships and railroad locomotives well into the 1880s, and the iron and steel industries only turned away from charcoal as their principal fuel in the 1860s. (In Britain by 1810 all iron furnaces burned coal or coke.) Thus the industrial revolution was accompanied not only by immense population expansion from immigration but by increased demand for lumber, fuelwood and wood-derived products. To give just one statistic, the amount of woodland cut over annually to keep the locomotives ahead of the train robbers in 1870 was 160–330,000 acres (64–133,000 hectares).

[12] M. Williams, *Americans and their Forests. A historical geography* (Cambridge University Press, Cambridge, 1989).

Thirdly, the pace of deforestation and especially the continued role of fire in the forests brought about a countervailing set of views. Indeed, in the development of these attitudes (especially perhaps in the west rather than the east) we can see the orgins of the US conservation movement, of which the wilderness idea as we know it today is an outgrowth. The conservation movement, with origins between 1870 and 1910, led to two interesting developments. These were, to begin with, the setting aside of large tracts of forest land under federal ownership (National Forests), which although not to be kept untouched were nevertheless to be managed so as to yield in perpetuity: to be subject to 'wise use'. The National Parks were a more extreme version of this ideology, starting at Yellowstone in 1872: total preservation was their aim. Then there was the discovery of fire suppression as a desirable and often achievable aim of forest management, whether on public or private forest lands. Poor forestry practices had added to the incidence of fires, since they left a thick deposit of inflammable slash on the ground which promoted high enough temperatures for the fire to run up trunks and start crown fires. So successful was forest fire suppression that in many areas of the western USA there is now the additional danger of crown fires because ground fires have been stopped before they could burn away the litter layer. Prescribed fire is now normal in the west and in the southern pine areas.

In spite of the creation of National Forests and National Parks in the Appalachian Mountains and in New England, there seem little doubt that none of these areas has been free from exploitation at one time or from excessive protection at another. However, some of them have now been under a form of conservation management for seventy years or more. Even when succession from open land to high forest takes 200–400 years, that is a sizeble chunk of time in which to be free from human direction of ecosystem processes. Add to that a series of cores of vegetation which may here and there have escaped thorough exploitation because of their site or remoteness, append also the sheer size of the forested areas in the eastern USA, and although perhaps not totally natural in status, a number of areas can be accorded the rank of runners-up.

The North American grassland

Even an untrained observer can usually see whether a woodland is recently planted, is actively managed, or just grown over, though the detection of 'ancient' status (meaning in Europe that there has been woodland on the site since at least medieval times) is more for specialists. But grasslands are a different matter, for unless a pasture is fenced and obviously subject to re-seeding and fertilizing, its historical status is not easy to spot at a glance. Cresting the Appalachians and coming out into the grasslands of North America, early European travellers were mostly convinced that what they saw was a natural phenomenon. So much so, indeed, that the grasslands of the Mississippi-Missouri basin were labelled 'The Great American desert', in spite of the fact that the presence of the Indians and some of their management practices was well recorded. Nevertheless the controversy about whether some, all or none of the grasslands of interior North America and the savannahs of the coastal south-east were natural or human-influenced continued well into the twentieth century. It is no surprise that the re-evaluation of fire as an ecological influence has been a significant element in the discussion of the origin of these grasslands.

The grasslands of North America, on the eve of European occupation, covered about 15 per cent of the land surface. There were various ecological and biogeographical types of grassland, dominated in the east by tall grasses such as the Big Bluestem (*Andropogon gerardi*), which can grow to 2–3 metres, and then by successively smaller species as the average precipitation declines westwards. In 1761, Charlevoix burst out of the woodlands of Appalachia to find 'immense prairies interspersed with small copses of wood . . . ; the grass is so very high that a man is lost among it. The words *pré* and *prairie*, of French origin deriving eventually from the Latin *pratum* or 'meadow', became adopted as the descriptive word for these areas of grassland in which there were at first small copses and then only riparian woodland. The word 'grassland' appeared only in the early nineteenth century, and became widespread in use only in the mid-twentieth century. Our theme centres around the nature of these grasslands in the period between the advent of colonizing Europeans from the east and the wholesale conversion of the grasslands to enclosed pasture and ploughland after 1870. Was this transition from woodland to grassland a climatic phenomenon? Was the passage from tall-grass to bunch-grass transition mostly controlled by the quantity and reliability of precipitation? Did large mammal herbivores such as the pronghorn antelope and the bison have a dominant effect on grassland species? Was lightning fire a control in vegetation development? Or, by contrast, was the vegetation largely a product of Indian management practices, analogous to those found in the eastern woodlands? Was their use of fire more important than natural lightning strikes? Did they exert a controlling influence upon the numbers and distribution of bison which in turn affected plant growth via the medium of grazing pressure?

In spite of considerable interest in these and related questions for many years now, and the challenge to scholars of re-appraising not only new evidence but ideological conceptions (about, for example, the ecological influence and environmental perceptions of the aboriginal inhabitants), the answers are not yet beyond debate. The role of human societies in bringing about changes in this environment is now accepted to be greater than scholars before 1945 would have believed; yet the overall influence of climatic cycles cannot be gainsaid. The answer presumably lies in the complex interaction of the two forces and in those details of history which cannot be now reached even after copious consumption of expensive continental lagers. The probability of some overall climatic control cannot simply be ignored. Rainfall does lessen westwards and becomes irregular from year to year in its incidence. Further, there seem to be cycles of perhaps eleven years in the amount of rainfall, which affect the biomass of the plants and hence possibly also the perception of the fertility by strangers. It is sometimes argued, for example, that the 1870s were a time of lush grass in the High Plains and this encouraged the view that their agricultural potential was high.

Also significant is the mass of data which show that grasslands tended to follow the Indians wherever they went. At the forest-grassland edge (the 'ecotone' in ecological terminology) they carried out the kind of woodland management encountered earlier in this chapter and this produced the grassland-with-copses type of landscape which Charlevoix, together with many other observers, described. The conversion of woodland to grass in this environment seems to have taken about three years, with annual fires as the key tool. The clinching evidence seems to be the fact that when the Indians were extirpated from this ecotone and permanent European settlement took place, the woodland became much more frequent and much more dense. The more intensive the development, the more dense the woodland, according to Pyne in his wide-ranging study of wildland fire in the USA.[13] Beyond the forest-prairie boundary, the Indian use of fire appears to have been very widespread. They used it for many purposes: to attract bison by providing an early bite in the spring from freshly burned areas; to direct bison herds by depriving them of grass; to drive the herds towards waiting bowmen, pitfall traps, cliffs or arroyos; to drive off mosquitoes, to bring up lizards which could be killed and eaten, to roast deer as it were *in situ* (an early example of the cultural trait which eventually led, one imagines, to the drive-in diner), and to harass human enemies both red and white. Early white travellers often reported seeing many Indian-set fires, and indeed were sometimes subjected to a persistent cloud of smoke brought to them from upwind as a signal of disapproval of their presence. Equally, the summer convectional storms produced lightning which set fire to the prairie: this is attested by many small-community newspaper reports from pioneer settlements as well as by the eyewitness of explorers.

Central to a discussion of the conversion of the plains vegetation is the ecology of the bison. The first bison to be seen by a white man was in 1521 at Anahuac in what is now Mexico, where Montezuma had a menagerie; in North America it was 1530 before one was recorded in southern Texas by Alvar Nuñez Cabeza de Vaca, which seems somehow appropriate. It quickly became clear that they were present on the plains in large numbers: one estimate suggests that in the 1700s there may have been 50 million animals, and herds which were 50 square miles (about 130 square kilometres) in extent, with half a million beasts in them, are mentioned in more than one account. Before they

[13] S. J. Pyne, *Fire in America. A cultural history of wildland and rural fire* (Princeton University Press, Princeton, 1982).

acquired the horse, the Indians slaughtered large numbers of bison by using fire drives, or by luring the animals to within stampeding distance of a cliff or canyon, although archaeological evidence suggests that the timing of the hunt often avoided the killing of gravid females. The spread of the horse northwards from the Spanish in what is now New Mexico started in about 1540 and the northernmost group (the Sarsi) had them by 1784. The combination of the horse and the rifle made bison-hunting much more effective, but no commentator seems to think that the numbers and hence the grassland ecology were greatly affected. The coming of the white man's culture and the railroads, however, reduced both the bison and the Indians to vestigial numbers, with 1876 signifying the demise of vitality in both components of the Plains ecosystem. This has raised the question whether the loss of bison made the Plains unduly lush, as previously grazed species of grasses began to assert more physiognomic dominance. This is very likely in some areas: there were more grass fires in Alberta in the late 1870s or early 1880s than previously and these are attributed to higher and denser vegetation. But the question is not very important since the diminution in bison was followed very quickly by agricultural use of the grasslands in the eastern areas and by cattle-grazing in the western. Once the range was fenced, these pastures inevitably became more and more semi-natural; with the advent of irrigation they are often now farmed intensively. In total, between 1833 and 1934 all the near-natural upland vegetation and two-thirds of the original marshlands had been transformed.[14] The savannah zone between forest and grassland saw, in general, an increase in the amount of secondary woodland, a trend which was still in place in the 1934–60 period.

So the question of the ecological standing of the North American grassland before the Europeans is still to some extent open: is it possible that fire was used by the aboriginal population right from the time of their first entry into this region? The vegetation of the Plains might then have been subject to natural fire, human-set fire and rainfall variability throughout its

[14] A. N. Auclair, 'Ecological factors in the development of intensive-management ecosystems in the mid-western United States', *Ecology*, 57 (1976), pp. 431–44.

Holocene history, with different influences being dominant from time to time and place to place. No single, all-encompassing verdict is possible. What is certain is that the early Europeans applied the cultural evaluation of desert and wilderness in no uncertain fashion, and wanted none of it.

The Mediterranean

The love-affair of northern Europeans with the Mediterranean has always contained an element of wonder at a quality of timelessness: many commentators have reached for phrases like 'time stands still' and 'just as it must have been when [Odysseus; St Paul; Petrarch; Jackie Onassis] stood here'. But there is no 'must', for a depth of human occupation and cultural change have engendered environmental changes, even if there have been longer periods of apparent stability than in some other regions.

We have rather patchy information about the economy and landscape relations of hunter-gatherers in the Mediterranean before the development early in the Holocene of agriculture in its various forms. During the last glaciations of northern Europe, the predominant vegetation in the Mediterranean seems to have been an unforested plain or steppe dominated by wormwoods and goosefoots (*Artemisia* spp. and Chenopodiaceae), with some oak woodlands. With the amelioration of climate there came more oak woodland, with juniper and pistachio admixtures. This woodland seems to have been open and in the glades grew many wild grasses, some of which were the ancestral species of cultivated cereals. In these surroundings, hunter-gatherers lived off many animal species, including the onager, wild ox, red deer, wild goat and gazelle. Among the important plant species was undoubtedly the oak, whose acorns were used as human food. Evidence of wear on some flint blades also shows that wild grasses were harvested, though whether as fodder for early domesticated animals or for human food is not known. At early Holocene sites in Palestine, there seems to have been a concentration on the gazelle as a source of meat, leading to speculation that attempts at domestication had been made. Hard evidence of permanent environmental manipulation by humans in the Upper Palaeolithic is sparse, but several scholars have extrapolated from the evidence of hearths in caves such as those on Mount Carmel to suggest landscape change. They point to the advantages of fire to hunters in the way animals are attracted to burned-over patches of ground since these often sprout plants quickly after fire, even before the autumn rains. Such areas, too, often grow grasses, bulbs and tuberous plants which might add to the food supply of human groups. So it seems probable that the long association of fire and mankind in the Mediterranean started in pre-agricultural times. The region is of course especially well suited to this relationship because of the long and dry summer which provides in most vegetation types a stock of fuel that can be ignited by thunderstorms or by human agency.

Hunting continued even when agriculture was established as the dominant way of life, as it still does. Some was for the pot and some simply for pleasure. In the Mediterranean, an historic form of hunting that had a strong impact upon ecosystems was the procurement of live animals for the Roman circuses. Since they had to be seen by thousands of spectators at once, only larger species were captured: elephants, ostriches, lions, leopards, hippos, tigers and crocodiles all featured at one time on the programme. With his talent for the unusual, Nero once put on a show of polar bears catching seals. At the dedication of the Colosseum, 9000 animals were destroyed in ten days; Trajan's conquest of Dacia was celebrated with the slaughter of 11,000 wild animals. It is scarcely surprising that the elephant, rhino and zebra were extirpated in North Africa, and lions in Thessaly, Syria and Asia Minor. This process would have interacted with land use change and the expansion of pastoralism, both inimical to the presence of wild animals, especially predators. Thus, as well as the enthusiastic spectators, it can be surmised that many inhabitants from those regions would not have objected; they still call for a renewed onslaught on the wolf in Spain today.

Of all regions of Eurasia, the Mediterranean was the first to receive agriculture. The basis of the pre-industrial agricultural system of the region was wheat, supplemented by a variety of tree and bush crops. Of these, olives were essential to human diet since they added to the fat content, as did a number of nut crops. Fruit was important in the diet, and the vine supplied carbohydrate as well as a highly tradeable commodity. This historic mix was much diversified by the Muslims, who brought sugar, citrus fruits and rice, as well as the carnation and the rose. The place of animals in Mediterranean agriculture is sometimes underestimated: they often formed an essential component of the diet and were if possible pastured near the village. If no forage was available, then long journeys might be undertaken to areas of summer pasture in mountain areas: transhumance from Provence to the Alps of Savoie for example. Transhumance inevitably spreads the environmental impact of grazing, which in the Mediterranean most often involved sheep and goats rather than cattle, which have daily water needs. Only in a few places could irrigated land be spared to grow fodder crops such as lucerne.

It is impossible to consider the pre-industrial ecosystems of the region without thinking of the towns and cities. In 1500, the four largest European cities, with populations between 100–200,000, included three in Mediterranean lands (Naples, Venice and Milan), two of which are actually ports on that sea. By 1600, Naples had grown out of that size bracket, but additions to it included Rome, Palermo and Messina (and also Seville, although the Guadalquivir reaches the sea west of the Straits of Gibraltar), and cities on the fringe of the region such as Constantinople and Cairo. Environmentally the cities as always transformed land to a condition of higher energy throughput, and acted as concentrators of materials. Thus they often had industries which necessitated the gathering of fuel from a large area or the diversion of watercourses to power mills. In turn some of these manufacturies produced wastes that contaminated rivers and estuaries: tanning, for example, is notorious in this respect. The urban population produced a demand for food most of which had to be met from local sources, although salted and smoked meat and fish might be traded long distances. Local land use was therefore intensified by such demands, though there might be a contrary trend in the retention of pasture for the many horses; the richer the city, the more horses. The rich, too, inevitably wanted their country houses and estates outside but not too far from the town, witness the villas to the north of Florence or inland from Venice. Competition for the land immediately outside the wall increased when new ways of fortification required a bare area or glacis to give the gunners a clear fire zone and a cover-bereft area to discourage sappers.

In spatial terms, nevertheless, the impact of agriculture and pastoralism dominated most of the pre-industrial landscape. Agriculture requires land transformation: the layout of fields is a simple example which may reflect as many differing social and political pressures as it does environmental influences. The Roman practice of centuriation, that is of laying out fields chess-board fashion in squares of 20 × 20 *actus* (710 metres square) still underlies the active landscape in parts of the north Italian plain, and is detectable in air photographs of the now near-desert of the Tunisian Sahel.

Figure 3.1 As an antidote to the usual pictures of the Mediterranean, this photograph of the Sahel zone of Tunisia shows the steppe-like condition produced by climatic change and pastoralism. This area was a wheat-producer in Roman times, but now will produce most crops (except olives) only with irrigation.
(*Photograph*: I. G. Simmons.)

That such a region should now be a near-desert raises interesting questions, and ones to which there is no totally accepted answer. To what extent, it might be asked, is climatic change responsible for the greater aridity of the landscape? Or is the driving force that of human activity in using the land so heavily first for agriculture and then pastoralism that only bare soil subject to windblow and gulley erosion from the winter rains is left? Many studies have tried to tease out these strands for the Late Roman and later periods, but few can reach conclusions which are sustainable over much time and space; it may have to be accepted that the interweaving is so subtle and pervasive that we cannot separate them out. There is a rather similar situation in studies of the causes of desertification in the world today.

Irrigation too has a long history. Certainly practised in Roman times, it was much extended in the sixteenth century as the towns grew. Rice could be added to the crop pattern and in effect a whole season (the summer) to the cropping year. In the lands of the Mediterranean, however, irrigation is necessarily confined since the areas of lowland and plain are limited. To cultivate the ubiquitous slopes requires another device: the terrace. Walls, mostly of dry-stone construction, are found on many slopes and provide a flexible resource: rainfed crops, irrigation and fodder can all be grown, and tiny pockets of production created out of impossibly rocky slopes. The opposite condition applies as well: there are places with too much water: in coastal marshes and lagoons, in river valleys liable to flood and in the great stretches of plain like the Po valley. All through the history of the region, attempts at drainage of such places were made; the early success of the Egyptians with clearing and irrigating the Nile lowland is well known. The Romans were highly active: the Pontine marshes south of Rome were tackled in 160 BC, and the Romans indeed started the long-term project of draining the Po valley, starting near Padua and Modena, producing land that the Italians called *bonifica*. As well as increasing crop land, such activities diminished the habitat of mosquitoes: malaria had been present at least in the eastern Mediterranean since 400 BC.

As elsewhere in the world, agriculture, pastoralism and the conversion of forests to open land produced soil erosion. Many Classical writers refer to this in rather stark terms, though it is by no means agreed that their words referred to all of Greece, nor that the kinds of soil loss that revealed the bare limestone bones of the landscape went on throughout history. Reforestation and careful forest management was almost certainly practised by some of the Italian republics: Venice, for example, looked carefully after the forests from which its fleets of galleys were built, not only in the Veneto but in Dalmatia and probably Crete as well. In this respect they were following in the footsteps of the Emperor Hadrian (AD 117–138), who had decreed the protection of forests in the Lebanon, Syria and Palestine. Metal-smelting and towns both required a sustainable supply of charcoal, which also led to woodland management rather than clearance.[15] Given, though, that there has been a long period of agriculture, and of pastoralism aided usually by fire, given the steep slopes surrounding much of the Mediterranean (especially the north coast), and adding the concentration of rainfall into a few months, then loss of soil and rock into the rivers, causing flooding in winter, and into the sea, allowing siltation of harbours and the creation of malarial lagoons, was inevitable.

Not all land transformation in the Mediterranean was in the service of food-getting. Quarrying might be localized where there was good stone, but would then exert a very strong environmental impact; Mount Pentelius bore a gleaming white scar where a particular marble had been taken out. For hundreds of years, stone from the small island of Brac off the Dalmatian coast was sought for buildings in the northern Adriatic region. Mining for silver, lead, copper, mercury and arsenic have all been carried out around the Mediterranean, with Spain an especially rich source. Surface and ground waters would then become clogged with debris from the mines and the run-off almost certainly become laden with toxic substances. In the search for gold in Andalucia in Roman times, hushing (the temporary damming of a stream and then the sudden release of the head of water over the ground to expose the mineral vein) was used, just as it was in the northern Pennines of England in the nineteenth century for lead. In history as now, pleasure was an important element of Mediterranean life.

[15] J. V. Thirgood, *Man and the Mediterranean Forest: a history of resource depletion* (Academic Press, London, 1981).

On the land we see for example the creation of beautiful gardens. The Egyptians may not have been the first, but their elaboration of rectangular enclosures containing water, shrubs, flowers and a summer house, and yielding fruit and fish as well as aesthetic satisfaction, set the tone for many later gardens. The formalized complexity of the gardens of Renaissance Italy, or the obvious conquest of the hydrological cycle in the fountain courts of the Alhambra at Granada, lie in that shadow. Utility and pleasure might also be combined, as in Classical Athens where the great philosophers often taught in a garden, as in Plato's Academe. Learning disappeared indoors as it moved northwards, first to the partially enclosed cloister and then to the windowless lecture-hall of today. The cloister reminds us that religion at various times would have protected environments from change, especially woodlands. The ancient Greeks, for instance, had sacred groves and the Mount Athos peninsula today maintains unmanaged woodland as a matrix for its secluded monasteries.

The lesson of these highly selected examples from the Mediterranean is this: even in 1800 there would have been few places that were in any sense natural. A few sand dunes, some mountain screes, cliffs, and perhaps the deeper parts of the sea itself that were little fished, would have been the only places that had not experienced change due directly or indirectly to human agency. That great Hellenophile, Lord Byron, put it a bit overdramatically when he said that 'man marks the earth with ruin', but we can imagine what he meant. It's as well he died in 1824.

As late as 1930, when the UK economy had only 6 per cent of its working population in primary industries, Spain and Italy retained 48 per cent, mostly in agriculture, though with some mineral extraction and fishing employment. Around the Mediterranean basin, the persistence of colonial regimes, including the highly conservative Ottoman Empire, retarded the development of modern industry still further in all except the north-western shore nations of Spain, Greece and Italy, with the last leading the transition to a true industrial society. Here we have good evidence that culture was the predominant element in postponing the development of a new way of life. There are no reasons of climate or natural resources which need necessarily have delayed industrialization. The dominance of colonial and imperial regimes was perhaps important in making sure that most

Mediterranean lands provided manpower and materials but were not contaminated with new ideas like freedom and democracy, which often seemed to go with factories and growing towns. Why do so many towns and cities from Paris southwards have wide boulevards in those parts developed or redeveloped in the nineteenth century? Civic pride, maybe, but they make it very difficult for insurrectionists to throw up barricades. Of the three states which made some progress towards industrialization, Greece remained the most traditional in its economy, with some 10 per cent more people in its primary sector than even Italy and Spain. In fact, it could be said that until the First World War, Greece experienced very little of the industrial revolution, one sign being a constant national rate of population growth, which was little regionalized. There were a few exceptions, like the town of Volos in Thessaly, which grew from a population of 5000 in 1880 to 25,000 in 1907. The main factor there was the opening of three factories, two of which made agricultural machinery while the third processed tobacco, so that the link with the agricultural phase of cultural ecology remained strong.

Of Spain, one economic historian has written of the 'failure' of the industrial revolution between 1830 and 1914. Through this period, the major evidence of it was probably the extensive working of minerals, mostly for export: lead, iron, zinc, mercury and coal were all produced in some quantity from the south-east Sierra, Murcia, Almeria, Vizcaya, Asturias and Cordoba. At one time it looked as though Andalucia might become a fully industrialized province, but this prospect had vanished by 1880 and the region became effectively de-industrialized. The exceptional area to this story is Catalonia, especially the area around Barcelona. Here, a native cotton industry provided the capital to develop chemical and metal industries, and produced a static steam engine in 1849, a railway locomotive in 1854 and iron ships by 1857. Further industrialization of this region, with the usual environmental changes from the change of economy, came with the construction of superphosphate plants in the Huelva-Barcelona region in the late nineteenth century.

For a fuller penetration of industrialism into economy and environment we need to turn to Italy, and especially to the north. As cotton in Spain provided a link between a solar-based economy and a later coal-fired one, so in Italy the silk industry provides an

Figure 3.2 Mountains – as here in the Austrian Tyrol – are popular for recreation and even in summer the swathe cut through the forest for the chair-lift is clear. Skiers change the vegetation by slicing off regenerating trees and by compacting the snow so that its insulating qualities are diminished. If artificial snow is made, the consumption of electricity for that purpose will probably exceed that of the settlement at which the skiers are staying.
(*Photograph*: I. G. Simmons.)

analogous connection. From the sixteenth century, the demand for silk had led to the cultivation of the mulberry tree throughout Mediterranean France and Italy, and by the mid-nineteenth century there were 600 silk-throwing mills in northern Italy, mostly using water power. The industry then turned to silk weaving as well, and to other forms of power, and in doing so became the nucleus for other kinds of factory production. There was, however, no iron and steel industry until the 1880s, when Genoa and Terni (Umbria) began production, and stimulated further demand for coal and iron ore in places like Fiume and the Val d'Aosta, with consequent changes to the condition of the atmosphere and the rivers. Machine industries began to develop, mostly centred in established cities like Genoa, Naples and Milan, and then the transport industry opened with bicycles in Milan and Padua, cars in Turin (Fiat) and Milan (Bugatti) and rubber also in Milan (Pirelli in 1872). The auto industry in Italy at this time, however, was definitely a small-scale luxury-market affair. The fast years of growth were 1897–1913, and one environmental consequence arose from the lack of cheap domestic coal: this was the development of the Alpine rivers for hydropower. Italy was the first nation ever to transmit hydroelectric power (HEP) into a city, with a 5 kV line between Tivoli and Rome in 1892. Before that, HEP power had been declared a public property in 1865, which led immediately to plants on the Adda and Ticino rivers, with larger plants coming in the last decade of the century. All of this can perhaps be symbolized by the fact that La Scala opera house in Milan was the first of its kind to be lit electrically.

Around the north-western Mediterranean in 1850 there were five cities with an industrial economy of a distinctly nineteenth-century kind: Barcelona, Marseilles, Milan, Rome and Naples. By 1914, it would be possible to talk as well of a number of industrial concentrations which not only affected their local environments but caused change in their outreach for energy and materials as well. These were Barcelona, Marseilles, Genoa, Milan, Turin, Verona-Padova, and Livorno-Pisa-Firenze. In Italy, though, textiles were still very important in most of these, emphasizing that the links with the earlier period were still in place. Italy was still somewhat sheltered from the brisk changes to the north: in 1870 there were only two rail routes through the Alps to her territory, but in the next forty years another ten were added. Among other things, Italy became opened up to the more northern middle-class culture-seekers among whom the English have always featured prominently. Such visitors (at any rate, the ones who would have passed Miss Lavish's examination paper at Dover in E. M. Forster's *A Room with a View*) are those whom Catherine Delano Smith describes as being sufficiently well-versed in classical literature and art as to find scenes 'too familiar to question' in Mediterranean Europe in the second half of the nineteenth century.[16] This may give us the clue that human-induced environmental change in the Mediterranean basin is subject to long periods of stability, or at least of slow and barely perceptible change. On the other hand, there have been large-scale transformations of great river valleys that have proved permanent, such as that of the Nile or the reclamation of the marismas of the Po. Not all such changes were permanent. Many of the Roman wheatlands of North Africa are now semi-arid steppe, with only the remnants of their centuriation pattern to remind us of their one-time fertility. The great coniferous forests of the region, in the mountains of the Atlas or in the Levant, seem to have gone for ever, leaving the cedars of Lebanon as fragmented as the politics of their homeland. At times, the steady attrition was accelerated: after the Reconquest, in the late thirteenth century, the deforestation of Spain speeded up. In the seventeenth and eighteenth centuries we have the example of the alpine valleys of Provence being cleared for viticulture. In the early nineteenth century the whole basin experienced woodland loss on a hitherto unimagined scale as population expanded. Every rural inhabitant needed 1.5 kilogrammes of dry wood per day for fuel alone, apart from other uses.

The vegetation type we now associate most with the region is the low scrub of oak and juniper, interspersed with grassland or with tufts of flowers. This is the great product of fire upon the former forests, the French *garrigue* or *maquis*, *phrygana* in Greek, *tomillares* in Spanish. From antiquity, Vergil recorded 'scattered fires, set by the shepherds in the woods, when the wind is right', and the Arabs carried on this practice for at least another

[16] C. Delano Smith, *Western Mediterranean Europe. A historical geography of Italy, Spain and southern France since the Neolithic* (Academic Press, London, 1979).

thousand years. Once created, the *maquis* is very dense unless subjected to more fire, and so the shepherds carried on burning to produce the vegetation they wanted. Where frequency of fire and density of animals has been high, the result is a low sward of asphodels, brilliant in spring but soon burned off by the sun. Today, this vegetation type accounts for 15 per cent of the land cover of Greece, for example, and some 10,000 hectares per year of it are still subject to fire. The *maquis* can stand as a symbol for the continuity of the use of fire as a tool of environmental management (and accidental change) from the Upper Palaeolithic to the present.

If we accept, then, the lack of any dramatic or revolutionary change, it is no surprise that the landscape elements of *agri*, *saltus* and *silva* (fields, pasture and woods) have an air of timelessness. If we add to the landscape the presence of the town, then we have the basis for the aesthetic appeal of this environment to northerners for several centuries. Further, it was the place in which the painting of landscape was first practised, thus bringing the past into close proximity with any present; again, it is a landscape which bares its past rather than burying it under two metres of brown-earth soils. So the geographer J. M. Houston can eloquently write: 'Mediterranean landscapes have the enduring qualities of an eternal present, like beaches on which the tides of successive civilisations have heaped their assorted legacies.' Given the perceived value of the paintings, and the sense of the past made visible in the landscape, it is not surprising that so many people see this region's environment, even today (and certainly between the coming of the railways and 1939), as evidence for the existence of a Golden Age of mankind. No matter, perhaps, that such a stage never existed: the regional relations between humans and environment was such that one group of outsiders was able to identify it with one of the most enduring myths that mankind has ever produced.

The oceans

Lord Byron, quoted above, went on to aver that the human-caused ruin of which he wrote stopped with the shore and that ten thousand fleets would not stop the blue oceans rolling on. Indeed not, but in dealing with the Earth's biggest region it is certainly the case that human activity has affected the ecology of the oceans. Most components have been affected in the coastal and inshore areas: plants and animals, salinity, seabed physiognomy, water chemistry and temperature, the content of silt and other solids form a partial list.

Of most interest, perhaps, is the ability of mankind to affect fish and mammal populations, since we might think that the very volume of the seas, along with their lack of transparency to the human eye, provided enough cover (especially in the open oceans) for fish, seals and whales to feed and reproduce abundantly, with any resource use by humans being a mere fraction of nature's bounty. In 1883, at an international Fisheries Exhibition in London, Professor T. H. Huxley (then President of the Royal Society) confidently asserted that at sea,

> the multitudes of these fishes is so inconceivably great that the number we catch is relatively insignificant; and, secondly, that the destruction effected by the fishermen cannot sensibly increase the death rate . . .
>
> I believe, then, that the [cod, herring, pilchard, mackerel] fishery and probably all the great sea fisheries are inexhaustible; that is to say, nothing we do seriously affects the numbers of the fish.

Yet only ten years later, a Select Committee of the House of Commons recommended (though no action was in fact taken) the adoption of minimum landing sizes for flatfish, the extension of fishing limits, and the establishment of a Fisheries Board, one of whose aims was to be the collection of better statistics. All of which shows (a) that there was considerable change for the worse in the UK fishery's ecosystems, and (b), more encouragingly, that Professors FRS may be wrong.

What had happened was, of course, the beginnings of the industrialization of fishing around British shores, a process which was pioneered there but which spread soon to most other industrializing nations and which has eventually penetrated all but the most conservative artisanal fishing enterprises of the globe. The biological context of 'the destruction effected by the fishermen' has always been that many fisheries operate in relatively shallow waters, that many fish either lie on the sea-floor or swim in dense shoals, that sea-mammals are ungainly on land and often have slow reproduction rates, and that many natural populations are cyclic in abundance.[17]

Even before industrialization, heavy densities of fishing effort were deployed in basins like the North Sea. The year 1828 saw the start of the deployment of fleets of smacks. These stayed at sea for several weeks and fed larger boats which carried the fish to a port. The catch was ferried between smack and carrier by small boats in charge of boys: 200 of these died in the North Sea between June 1880 and December 1892, and of the 4103 who became apprentices in the fleets in that period, 1080 absconded. (Before 1880 a death at sea did not have to be reported unless the boat were damaged.) To this scale of fishing effort was added the power of steam when in 1860 a paddle-steamer tug from Sunderland, the *Heatherbelle*, towed two smacks 8–16 kilometres offshore; by 1864 there were twenty-four such rigs working off the north-east coast. The fishermen were now independent of wind and tide but, almost as important, the steam winch could haul up bigger and fuller nets from greater depths. Thereafter, no fishery was immune from heavy impact and a map of the occurrence of depleted fish stocks through to 1914 is a map of the extension of industrialized fisheries catering to the demand for protein in the rapidly increasing populations of the newly industrialized countries and aided by demand-extending technologies such as canning and railway transport, the on-shore construction of large ice-plants and eventually ship-borne refrigeration. The core of modern fishery practice was the North Sea, the North Atlantic between Cape Hatteras and the St Lawrence, and between Seattle and the Aleutians. In the first-named of these, the catch per unit effort declined by about one half in 1889–98, and in the third (the Pacific halibut fishery) the stock density fell by a factor of seven between 1910 and 1931. This sent the modern steam trawlers (whose forebears date from 1881) out into the deepest oceans to search for yet more fish stocks.

[17] D. H. Cushing, *The Provident Sea* (Cambridge University Press, Cambridge, 1988).

Yet even in remote corners of the world oceans, fishermen found that they were not the first exploiters of the marine resources. In his voyage of 1772–5 Captain Cook encountered the southern polar continent and its islands and later suggested that 'the world will derive no benefit' from the region. Nevertheless, between 1778 and 1830 the Southern Fur Seal of South Georgia was brought to the verge of extinction, a story repeated for many other populations of fur seals. One side-effect of the great explorations of the oceans and their margins was the revelation of new whale stocks well away from the commonly utilized areas such as Biscay, Spitsbergen, New England and the Davis Strait. Many other zones of the oceans thus came to be worked by long-distance whalers from Europe and North America before the 1880s. The industrialization of this resource process depended like the fisheries upon steam vessels, but can also be laid at the feet of one Sven Føyn, who in 1873 patented the explosive harpoon. By 1920 attention had been drawn to collapses in the stocks of at least six whale populations, and the subsequent story is well known. 'The great immeasurable sea' of the psalmist now contains so relatively few whales that they can, just about, be numbered.

On the margins

We are apt to concentrate on the big and important regions and issues: those which provide today's crops and living areas or which are constantly brought to our attention like the marine mammals. We might remember, though, that there are less visible parts of the world which have also undergone rapid change at human hands.

The case of the very high mountains is instructive, since the history of environmental change at human hands provides an insight into today's development problems, especially in the Himalaya. Here a sequence of environmental changes, apparently human-induced, has led to a whole theory of mountain degradation, of which the main elements can be summarized as:

1 Post-1945 population growth

2 Subsequent increased demand for fuelwood, fodder and cropland

3 Massive deforestation in which 50 per cent of Nepal's forest reserves disappeared in 1950–80

4 This deforestation leads to massive soil erosion and accelerated incidence of landslides

5 More deforestation and more terrace-building for cropland means further for the women to go for wood and so eventually more dung is used for fuel and the fertility of the cropland declines

6 The new land use pattern means increased run-off during the summer monsoon and so we find:

   (a) Flooding and siltation in the plains south of the Himalaya
   (b) In the hills, lower water levels in streams and wells during the dry season
   (c) Silting up of reservoirs
   (d) Sand and gravel spreads across agricultural land in the plains
   (e) Formation of islands and sandbanks in the Bay of Bengal from the silt loads of the Ganges and Brahmaputra.

But what is shown by a detailed area-by-area study of the environmental history of the Himalaya, where the evidence is available, is that deforestation was often intensive after the 1770s and climaxed in the period 1890–1930. Thereafter, oral history and some repeat photography (that is, photography of the same scene at later intervals) shows secondary forest. For some regions, therefore, there is a 400-year history of environmental change which contradicts the current theory and throws new light on development efforts which assume all the linkages implicit in 1–6 above. This shows, *inter alia*, how unwise it may be to extrapolate the occurrence of current environmental processes (a) over large areas where the evidence is unexamined and (b) into the past, without constructing a detailed environmental history.

One of the reasons why it is possible easily to absorb a theory such as that described for the Himalaya is the ecological/land capability classification of mountains as lacking ecological resilience, that is of being fragile environments. One other such set of ecosystems is that of islands, particularly oceanic islands. These often have a flora and fauna which lacks species variety since natural dispersal to these remote places was not easy; isolation has produced many endemic species so that species loss quickly notches up the demise of rare plants and animals; and remoteness may mean that many species have failed to evolve defences against particular classes of predators, so that flightless birds, for example, are commonest on oceanic islands.

Into such environments have come various human groups. Some, as in the South Pacific, came to settle and practise agriculture and produced habitat change as well as pressure on indigenous animal populations; others, where visited by European sailors, produced inadvertent change by landing pigs and goats to breed and provide fresh meat for the return journey, or by killing endemic species for food rendered their populations unviable (the dodo and the great auk are classic examples), or simply by providing the rats with an opportunity for a run ashore so that they missed the sailing time: heaven for rats includes a dense population of year-round ground-nesting birds.

A well-documented example is that of the Hawaiian Islands. Occupied by Polynesians in about AD 400, they were 'discovered' by Europeans in 1778 and various ecological changes followed both events. In this case, the Polynesians burned and cleared much of the lowlands for settlement and agriculture and introduced dogs, chickens, pigs and Pacific rats as well as non-native plants such as kukui (*Aleurites exulans*). In 1778 Captain Cook brought European strains of rats, pig and goats, and fifteen years later George Vancouver avenged Cook's death by landing cattle and sheep. Goats and cattle have probably caused the most change in the flora and fauna of Hawaii, the latter partly because a taboo on killing them existed until the dominance of American values was established and so feral populations of both goats and cattle became established. The resulting changes have been of two types: direct habitat destruction, and displacement of native species by introduced species, both wild and domestic. These had a great impact upon a flora in which 96 per cent of the flowering plants were endemic species, as were nearly all the forest birds. At least 39 taxa of birds became extinct before 1778 (including 13–15 flightless species) and 28 taxa in the last 200 years, and 53 taxa of plants and animals are now designated as 'endangered' or 'threatened', including an entire endemic genus of tree snails (*Achatinella* spp.) and familiar birds such as the 'o'u (*Psittirostra psittacea*) of which there are only 300 left and in whose population avian malaria is widespread. The diminished number of native species and habitats has led to a vigorous conservation programme led by the state government of the islands, though there are powerful influences at work for continued developments of the kind that will lead to further species loss.

A similar set of processes is also writ large in the most famous island group in the biological world, the Galápagos. This archipelago was discovered by Europeans in 1535 and became a home to pirates, whalers, hunters, cultivators and fruit farmers. Darwin on his visit in 1835 noted that the inhabitants' diet of bananas and sweet potatoes was supplemented by feral pigs and goats and that two days' hunting of giant tortoises supplied enough meat for the rest of the week. By 1887 Floreana had orange and lemon trees and wild donkeys. Today, the vegetation and hence the whole ecology is influenced by feral populations of goats, donkeys, pigs, horses, cattle, rats and mice. There are also over 200 introduced plant species. The remarkable native fauna of these islands, together with their perceived role in the history of science, clearly requires very careful as well as strict management if they are to be perpetuated.

At a distance of 3872 kilometres to the south-south-west lies Easter Island, and palaeoecological work has thrown light on the enigmatic cultural history of that remote triangle of lava, 3747 kilometres from South America, the nearest land mass. In spite of Pleistocene and recent climatic changes, the island remained forested until the period AD 750–1150. Then there are rapid falls in the quantity of trees, with a final surge of deforestation at about AD 1450. These dates coincide with no climatic changes, but are closer to the arrival of humans (and rats) about AD 450, and eventually to the demise of the culture responsible for the famous megalithic statues, *c.* AD 1680. Population growth on an isolated island seems to have led to deforestation and this in turn removed the trees necessary for moving the statues as well as for building canoes for fishing. Food shortages and warfare are both well documented, and there has been eventual population decline.[18] So environmental change at human hands is strongly implicated in the demise of a remarkable cultural trait – though not, it seems, in explaining why the practice of constructing megaliths flourished so in the first place.

Another kind of marginal land-form is that of wetlands, where the substrate is subject either to periodic (diurnal, seasonal, unpredictable) or permanent inundation. Such lands are found at the margins of the oceans (for instance, mangroves), in estuaries (salt marshes and mud-flats), along river valleys (floodplain woodlands), around the edges of lakes (for example, reed-swamps) and in the form of bogs. Wetlands are found in many different climatic zones and with many different soil types and cover perhaps 6 per cent of the land surface of the globe.[19] However, they are more frequently distributed on small and local scales compared with forests and grasslands, for instance, and this encourages their conversion. The rates of conversion during various historical periods is less well known, however, than for drier habitats: in a much-studied nation like the USA, estimates suggest an 'original' (that is pre-contact) area of 50×10^6 square kilometres of wetland, with a present-day area of $21–37 \times 10^6$ square kilometres of such landscapes, but no closer appraisals have been made. Large and contiguous areas do, however, exist as with the great muskeg-forest zone of Canada, and the 25 per cent of the surface of Indonesia which is swamp. Transformation of these wetlands has taken place for a variety of purposes: where peaty soils were found, then an energy source was to hand. The Norfolk Broads, large areas of open water in eastern England (now a prized recreational and wildlife resource), and the Rijk den Duizend Eilandtjes in north Holland are examples of lakes which formed after the extraction of peat – in the former case, in medi-

[18] J. R. Flenley et al., 'The late Quaternary vegetational and climatic history of Easter Island', *Journal of Quaternary Science*, 6 (1991), pp. 85–115.

[19] M. Williams (ed.), *Wetlands. A threatened landscape* (IBG Special Publications 25, Blackwell, Oxford).

eval times. Other examples around the North Sea usually find agriculture to be the main beneficiary, and in developing nations today this is a major theme of the transformation of wetlands. A runner-up is often the production of animal protein from farmed fish, and so salt marshes and mangroves are turned into fishponds. Most wetland reclamation in the developing nations dates from the second half of this century, but in the now-industrial nations (and China) it has at least a 2000-year history.

Conversion for the many possible uses of transformed wetlands does not, however, completely remove one basic feature: the

Figure 3.3 The reclamation of wetlands has been popular for centuries around the North Sea. This is the East Fen of Lincolnshire, England, reclaimed in the early nineteenth century, but could be parts of the Fenland proper or of the Netherlands. Such areas shrink once dried out and have to be kept dry by pumps: hence they are now closely tied in to industrialism.
(*Photograph*: I. G. Simmons.)

presence at some time or other of a very high water-table; and so measures have often to be taken to protect the newly won resource. In the English fenland, drainage has meant that the peat has shrunk by 3 centimetres per year since 1848 and the combination of shrinkage, oxidation and fire can result in the lowering of land surfaces by 10 centimetres per year, with 50 centimetres possible in the first year after drainage. Hence, societies undertake an open-ended commitment to pumping of such regions, or to the maintenance of sea walls in coastal projects. In 1753, Frederick the Great boasted that a big drainage project in Prussia had enabled him to conquer a whole province in peacetime: unless it was an unusual undertaking, the initial battle would have had to be followed by a well-equipped army of occupation.

A variable palimpsest

The description of most landscapes except the little-affected as a palimpsest is by now something of a cliché. Few other images, however, convey so well the multiple layering which a panorama of a long-occupied land can give us. Just as a satellite view of the planet shows parts that are apparently little affected by humans as well as the made-over areas, so in many regions of the globe any panorama will show in the present landscape presences of past ecologies. For example, the river bank downstream from my house was covered in deciduous forest in prehistoric times. A fragment of this woodland has survived where it was part of the Prior of Durham's hunting park. Much of the woodland was converted to crop and grassland where the slope was gentle, and some of the medieval ridge-and-furrow survives to be seen in low light or snow. The industrial era intruded to the point of inserting a colliery into this landscape during the early nineteenth century. Its remains survive as a scrubby heap of ash and stones together with a sunken ring of turf where the shaft went down. Another coal pit was converted to a brickworks early this century; this too has gone, but its claypit and surrounding area is being reclaimed gradually by unmanaged deciduous woodland. So the tutored eye can see the marks on the land surface which give clues to a set of past environmental changes.

Many historic buildings can be seen not simply as isolated monuments but rather as symbols of a whole system of environmental management which may have provided lengthy stability, as with the monasteries of Mount Athos in Greece, or considerable change, as with the Cistercian monasteries of upland Britain. In the latter, the policies of animal husbandry would have changed the vegetation of the grazing lands on the uplands; bloomeries would have necessitated the management of woodland for a sustained yield of charcoal, and watercourses were diverted for domestic and quasi-industrial purposes. On Mount Athos, numbers of grazing animals were kept down by the prohibition on female beasts and so the woodland underwent very little conversion. We would expect a very high fire frequency in such terrain, but doubtless such an oasis of virtue was rewarded by a deflection of normal ecological processes.

Even in a terrain of rapid vegetation growth and soil formation, we can see some of the layers of the past; how much more must be buried just beneath the luxuriant growth of the humid tropics. Yet even more must have been lost from the world's more skeletal landscapes such as the African savannahs and the uplands surrounding the Mediterranean. What you see is usually only a small part of what you got, which was a strongly humanized landscape in most parts of the world even before the advent of industrialism. Many parts were much wilder than today, but few outside the most extreme environments and the open oceans were totally unaffected by humans and their practices.

4

'*Not dream of islands*': An Environmental History of Two Political Units

The history of the world as a whole and of particular biomes or other ecological units is of considerable interest in establishing the principles of environmental change using the customary methods of scientific investigation, whether these be of the experimental-predictive kind or the historically based,[1] together with the historical techniques of the humanities. Yet in the world of everyday life, these ecologically based units have little enough significance: it is not of great interest to the average voter in Japan, for example, to know that she is living in (the remains of) the temperate forest zone during an era of Pacific →→ Global shift in cultural ecology. What is more germane is the environmental history of Japan as it affects the life-ways and opportunities for resource use and for leisure of that individual, family or community, and its interaction with the most effective political decision-making levels for environmental matters, and especially those processes which are perceived as problems.

What follows, therefore, is an attempt to chronicle some of the environmental history of two political units as a background to today's situation, noting that not all this history is relevant to today: much of the traces of it have been altered by later natural and human-induced transformations and sometimes only hints of

[1] S. J. Gould, *Wonderful Life. The Burgess Shale and the nature of history* (Hutchinson, London, 1989).

it are present, and then only to the informed eye. But few nations are now indifferent to their history (of which their environmental history is at present usually a vital but unrecognized or at least undifferentiated part), if only for its value in the tourist trade. The main example I will describe is that of England and Wales: for part of the last 10,000 years of course they were separate kingdoms, but are now under the same legal system. (Much of what is said also applies to Scotland south of the Highlands, but the legal system is in some ways different. Ireland has had a separate ecology as well as a unique political history and will not be discussed here.) By way of comparison, an outline account is given for Japan for the agricultural and industrial periods. These accounts can be read alongside the other national narratives for the last 300 years given in B. L. Turner's immense compilation.[2]

England and Wales: towards an environmental history

These countries lie in the Temperate Forest biome: that is, if human activity were to cease suddenly and the resulting successions were to be telescoped in time, then the majority of the vegetation would be a mixed deciduous forest in which, we suppose, oaks, elm, beech and sycamore would be important trees. This forest would be different from the one that would be present had humans never existed, since sycamore is an introduced tree but one which is very successfully naturalized. This forest might, in today's climate, have an altitudinal limit at about 400 metres, beyond which the upland communities of moor and mire would take over, but we are not sure of the extent to which tree growth at those altitudes would be successful in the absence of fire and sheep. Almost certainly there would be a fringe of birch and rowan scrub beyond the upper specimens of oaks and pines. So much, though, is imagination, simply to show where we would be in the natural world, and indeed still misleadingly are in some

[2] B. L. Turner et al. (eds), *The Earth as Transformed by Human Action. Global and regional changes over the past 300 years* (Cambridge University Press, Cambridge, 1990).

atlas maps. The reality of our environment today is one of a mosaic of rural and urban land, inland waters and a much-amended coastline: the task of this narrative is to see how human communities and natural changes together have produced today's variegations. For easier linkage with other chapters, this account will be subdivided into the eras of cultural ecology described in chapter 1.

1 *The hunter-gatherer phase*

There were early groups of humans in Britain during the intervals of Pleistocene glaciation, including those in East Anglia during the Hoxnian discussed briefly in chapter 3. This account of the last 10,000 years, however, takes up where the last (so far) glacial phase ends and climatic amelioration proceeds quickly, accompanied by rising sea-level as glaciers and ice-sheets become liquid water once again. At *c.*8300 BC, the dominant culture of England and Wales is that of the Mesolithic, remains of which are found in both upland and lowland environments, and whose economy seems to have lasted until the firm establishment of agriculture in the period 3500–3000 BC.

The environment in which these people lived was at first driven largely by the forces of nature, in the sense that the land and seas surrounding Britain were responding quickly to the warming of climate. This brought about equally rapid changes in vegetation (though with a lag between temperatures and the establishment of the plant species less tolerant of cold) and a rapid though not smooth rise in sea-level, so that Ireland was separated off by 19,000 BC and the North Sea formed by about 7500 BC, with the last land remaining along a line from Hull to Esbjerg. Thereafter both humans and other organisms had to make a sea passage into Britain from the continent. The ecology of England and Wales immediately after the 8300 BC boundary that marks the last truly glacial episode (one largely confined to mountains) was dominated by tundra-type vegetation in which sedges, grasses, low heath plants and a great variety of the flowering plants of open habitats (nowadays confined mostly to mountains and unstable slopes) were dominant. In sheltered places the tree birches and the Scots Pine were found. Across these largely open landscapes would have been seen herds of mammal herbivores and their predators: the wild aurochs, wild horses and reindeer were especially frequent, along with the wolf. In wet lowlands, the moose was also present, and wherever woodlands flourished then deer were found. Streams were probably less flashy than now since many were held up by beaver dams. It would have seemed to our eyes a very wet landscape, since many stream-courses had fens with willow scrub and in the parts of the country that had been involved in the most recent glaciations there were many relict lakes which had not yet been filled in by silt and by organic debris from their plant life.

This tundra was relatively short-lived, since the warmth made possible the immigration of a succession of forest-forming trees such as hazel, elm, oak, lime and alder. (Beech and hornbeam followed later, during agricultural times.) These took over the tundra lands and formed a mixed deciduous forest over most of the dry lands of England and Wales. The exceptions were areas which were still very wet, such as small lakes and mires as for example in north Shropshire and south Cheshire, large areas of wetland such as the Somerset Levels and the fens of East Anglia, and the upland areas which are now mountain and moorland, such as the Lake District, the Pennines and North Wales. The tree-line on the uplands was, however, quite high in the warmest period of the post-glacial (*c.* 7500 BC) and high points like Cross Fell in the North Pennines (2930 ft/977 m) had trees on the summit. In western Wales the tree-line was rather lower in the face of strong western wind velocities.

In this rapidly changing ecology, the human groups lived as food collectors and not food producers. The archaeological evidence suggests that most species of mammal herbivores and fish were taken as prey by groups which lived in all the different habitats at one or other time of year. The only evidence of settlement type comes from lowland wetlands, which may not be typical of the entire economy-environment relations: small groups camped at the edge of reed-swamp in the midst of birch woodland seem to be the pattern. Evidence of tools in the uplands suggest that these were occupied as well. The most interesting feature of the forest period of Mesolithic time is the extent to which the human groups may have changed their environment, rather than simply having been children of nature. Evidence has accumulated that during the forest phase of the Mesolithic, human communities attempted to manipulate vegetation for economic ends, as discussed briefly in chapters 1 and 3. The creation of small openings in the deciduous woodland, or perhaps the maintenance of the naturally occurring openings, would have been to the economic advantage of the human groups: the openings would have attracted mammal herbivores to the grassy floor and the marginal browse. The replacement of high forest with more shrubby vegetation such as hazel would have also yielded a crop of nuts with high nutritional value. One lasting result of forest removal was the growth of peat in water-retentive areas such as stream beds and in water-shedding areas with a relatively low slope.

So the hunter-gatherer period in the history of England and Wales ends with an environment which was still largely natural, but which was in some places being converted to near-natural ecosystems by human activity. Not all these conversions were permanent, but enough of the changes were sufficiently long-lived to provide evidence that the humanization of the environment was beginning. In effect the story of the next two eras is that of the differentiation of the forest matrix into a number of distinct habitats and communities, some of which retain traces of their natural origin, while others of which are very largely human creations.[3]

2 *The agricultural phase*

Agriculture began to be practised in England and Wales from about 3500 BC, and was superseded by industry as the major source of production and livelihood from the middle of the twentieth century. In both the agricultural and industrial eras, England and Wales were the heart of an empire whose economic and political linkages had its feedback into the environmental relations of the mother countries.

It is not known whether agriculture entered England and Wales

[3] S. R. J. Woodell (ed.), *The English Landscape, Past, Present and Future* (Oxford University Press, Oxford, 1985).

in the shape of a technology which was adopted by indigenous hunter-gatherers or with an immigrant population bearing the new combination of cereals and cattle. Either group, however, stood to benefit from the environment created by the last hunters, with a patchwork of openings, scrub and high forest, for into such a mosaic the small-scale agriculture would easily fit. It was once thought that this early agriculture was of the swiddening kind ('slash and burn'), but opinion is swinging at present more in favour of small but permanently cultivated patches. Whatever the actual pattern, there seems to have been, in the north at least, some cessation of the use of fire as a management tool, since in the contemporary deposits the frequency of charcoal declines with the onset of agriculture and in some places trees grew back over soils of moor and heath.

There is clear evidence from Bronze Age times (*c.*1800 BC) that there were permanent fields. It comes mostly from uplands like Dartmoor and North Yorkshire, probably because these uplands preserve the evidence the best. The landscape was divided up into rectilinear plots and some of these were most probably used for cereals and others as corrals for domestic beasts. The presence of linear dykes and tumuli on ridges suggest that estates were formed with a core of cultivated land and a periphery of extensive grazing and woodland; some parish boundaries still follow the outlines of these prehistoric events. One feature is, environmentally, important: a number of the fields show a scattering of pot fragments, which allows the interpretation that the settlements' midden-heaps were spread on them. Thus these fields are linked with the general feature of pre-industrial agriculture in so many places, that is that the upkeep of fertility levels in the soils was a key feature. In this, the cattle (and other domesticates such as pigs and sheep) were very important, since they acted as sources of manure, turning vegetable matter such as grass and tree leaves into dung. The discovery that pulses such as peas and beans were nitrogen-fixers in soils and that they could be rotated with the cereals which were all-important for human nutrition was a stabilizing factor in agricultural output.

Historically inept though the aggregation may seem, there is hence one main story for the environmental relations of agriculture in England and Wales for perhaps 4000 years: the need to

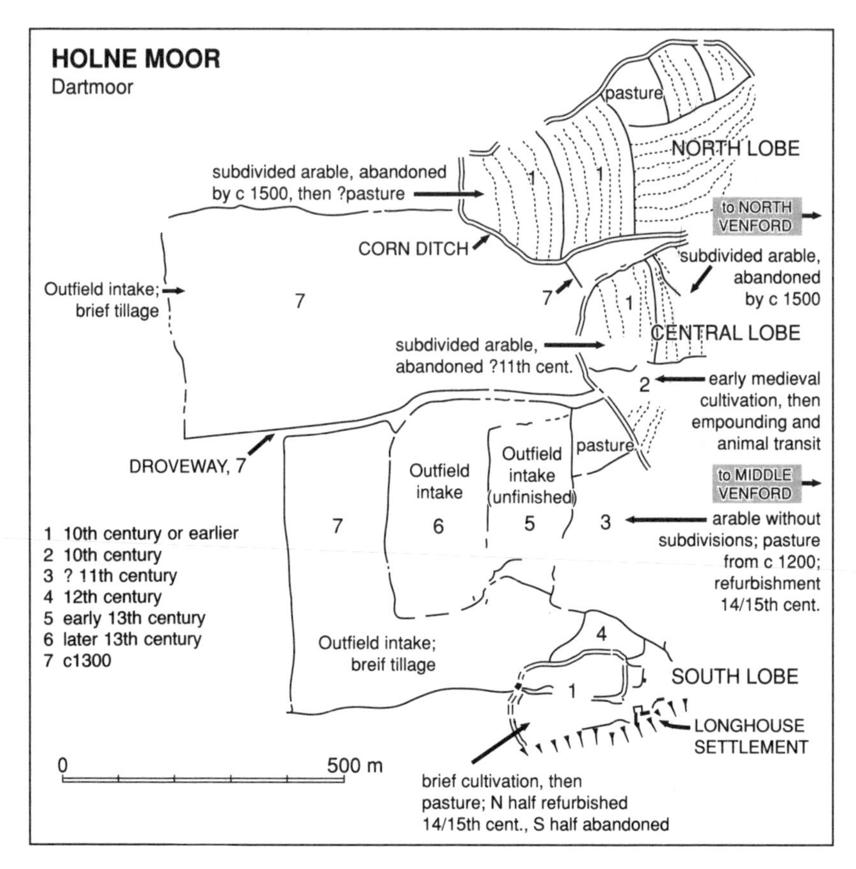

Figure 4.1 A settlement on the edge of Dartmoor, showing the changing intake and abandonment pattern of fields and uncultivated land.
(*Source*: G. Astill and A. Grant (eds), *The Countryside of Medieval England*, Blackwell, Oxford, 1988, fig. 4.5.)

keep up the fertility of the soils and the parallel demand either to intensify production or to extend the cropped area as human populations grew. The response to these imperatives changed down the years but their basic nature stayed the same until either massive imports became possible or production could be subsidized with fossil fuels.

Intensification was always helped by improved ploughs. The introduction of metal tips to the coulter assisted the cultivation of soils hitherto resistant to anything other than scratching; the provision of a mould-board and later an all-metal mould-board, the replacement of the ox by the horse and the spread of the horse-collar as a way of harnessing the motive power of a horse all helped production by aerating the soil, by bringing up fresh nutrients from weathered subsoil and by burying weeds. One result, however, was accelerated soil loss and this contributed to silting of rivers and the deposition of terrace deposits of alluvium in valleys; occasionally also to thick spreads of colluvium on valley sides, echoing (though rather faintly) similar processes in the Classical Mediterranean. Energy spent on fencing and bird-scaring was repaid in lower pre-harvest losses to birds and mammals; better ventilated storage meant less post-harvest loss to fungi and more secure storage; semi-domesticated cats helped reduce the rat population, as indeed did the provision of owl-access holes in barns and byres.

In England and Wales, higher productivity seems always to have been associated with discretely owned fields and so the enclosure of open fields, of which there was a burst in the sixteenth and seventeenth centuries can be seen in that light: investment in higher fertility levels is more attractive when you or your descendants will inherit the benefits. Indeed, many classes of environmental resources are more sustainable when in private ownership rather than as part of a commons, a lesson still being applied to sea fisheries, for instance. Enclosure on the grand scale led to the formation of great estates, some of which pioneered fertility-maintaining innovations like four-course rotations and production-enhancers like the seed drill. Short leys and turnips might replace fallow and thus provide food for more animals, improving nutrition and the quality of manure all at the same time. Tapping hitherto unused sources of vegetation through new introductions can be seen as the rationale for the spread of the rabbit, brought in from the continent in the twelfth century (the first records from the Isles of Scilly) and much valued for its production of meat and skins. Warrening might take large areas of land which was cropped to the bare soil; elsewhere uncontrolled rabbits were a pest: as early as the thirteenth century, wheat in Sussex was said to have been devoured year after year by the rabbits belonging to the Bishop of Chichester. The area of grassland within the system might vary with both environment and economy: there seem to have been falls in the proportion of arable land from nearly 100 per cent to just over 50 per cent between 1300 and 1500 in the Midlands and from a proportionately lower maximum elsewhere.[4] Demand for meat was restrained by a religious ban on the consumption of four-legged animals for up to three days a week, and throughout Lent and after 732 Pope Gregory II banned eating the horse entirely, a prohibition not now entirely observed throughout his part of Christendom.

To extend the agricultural area in prehistoric through medieval England and Wales meant turning initially to the forests. These acted as a kind of land bank, in spite of their other great supplies of renewable environmental resources. So at periods of population growth, the forests are cleared for cultivation, for example in the twelfth to fourteenth centuries; when population declines as at the Black Death, the woods return again. The environmental consequences are considerable: the soils are open to physical and chemical loss and the materials have to go somewhere. Quite deep terraces of silt along the Midland rivers of England testify to Bronze Age clearance, for example, and we cannot doubt that the nutrient loads caused eutrophication of the fresh waters. Loss of woodland area meant that the remaining forests had to be managed more intensively for sustained yields.

As in the riverine civilizations discussed in chapter 2, the cropland core might have a surrounding area of pastoralist economies. This is most marked where there are uplands which by medieval times had emerged as moorlands, and in areas of heathland and grassland down. Sheep and cattle were the favoured animals, but most were depastured daily (so that some at least of the manure might be collected) or even outwintered once the medieval period was past; before that, though, there might be a summer journey to upland pastures, as testified by the place-names containing elements akin to 'shieling' and 'saeter', of which 'scale' is the commonest survivor, as found in Cumbria, for instance; the practice survived until the seventeenth century in the Border lands. Some shielings in the northern Marches were converted to permanent settlements in the later twelfth and thir-

[4] G. Astill and A. Grant (eds), *The Countryside of Medieval England* (Blackwell, Oxford, 1988).

Figure 4.2 Beating an oak tree for acorns to feed pigs. Note that the tree has been shredded, i.e. all the side branches have been removed, probably for leaf fodder for cattle. In the background, it looks as if other trees have been coppiced. This is in fact French, but might have been almost anywhere in medieval Europe. (*Source*: Occupations of the Months, from the Calendar of the Playfair Book of Hours, French, late fifteenth century. Reproduced by courtesy of the Board of Trustees of the Victoria & Albert Museum, London.)

teenth centuries only to be abandoned when various combinations of livestock plagues, crop failures, Scots raiders and the Black Death made them untenable.[5]

To keep their production high in the face of a growing season for plants of only weeks, graziers used fire to burn off any old vegetation which might shade new growth in early spring. Heather was thus discouraged at the expense of cotton-grass on peaty sites, since the latter put out fresh greenery for an early 'bite' for sheep; areas vegetated with purple moor grass were burned regularly, since it is deciduous and firing releases nutrients for new growth

[5] A. J. L. Winchester, *Landscape and Society in Medieval Cumbria* (John Donald, Edinburgh, 1987).

as well as removing inedible and shading litter. A major result of burning and grazing was the keeping open of the heaths and moors: in the face of fire and teeth, no trees can regenerate, though a scrub will quickly form if these practices cease. The Statute of Merton in 1235 confirmed the soil rights of the Lord of the Manor even on common land, but it also gave the inhabitants of surounding vills economic rights, and it was in their interests to keep up the pressure of animals upon the land. One result is that palatable species have become less common in the vegetation and inedible species such as bracken (which is toxic as well) have flourished as a consequence. Much of this spread is, however, post-medieval; in earlier times bracken 'rooms' were prized for their yield of thatching materials, cattle bedding and ash for soap.

Learning how best to use moorlands for animal rearing made the expansion of cattle into the uplands of northern England highly profitable by the early fourteenth century: the upper valleys and high moors near, for instance, Skipton and Barnard Castle supported large herds, with hundreds of animals: they were indeed cattle country, and the transition to the more recent emphasis on sheep may be interpreted in part as a response to the changes in vegetation brought about by those dense herds, to the point where the remaining grasses provided fodder for sheep but no longer for cattle: watch them eat, and see the difference between the tongue-and-pull action of a cow and the lawn-mower nibble of a ewe. Cattle also provide material for tanneries; these were often in towns, but in the Derwent fells of the Lake District in 1278 there were eight tanneries, whose demand for oak bark ensured attention to woodland conservation, though not to its extension.

If extra food is required but there are no radically new intensifications available, then either a bigger area or imports are the usual responses. To achieve the first, recourse is had to areas of less intensively used land which can be converted. So the moorland edge, areas of heathland on sandy soils as in the London basin, Dorset or Suffolk, salt marshes and estuarine mud-flats were all at some time considered for being fenced or embanked, ploughed, manured and sown. At times of very high prices, such areas might be taken in and then allowed to revert quite quickly with a 'catch crop' of barley or wheat being possible for one or two years. The story of the conversion of the Fenlands of East Anglia during the seventeenth century from a wilderness of reed-swamp, willow thicket and open pools to a series of large rectangular fields growing cereals (and needing continuous pumping to keep them dry enough to do so) is well known. The Somerset Levels underwent similar changes in the eighteenth century, save that permanent grassland was the commonest contemporary outcome rather than arable land. The environmental consequences of these actions is also apparent: the diverse fauna of the Fens is reduced to the minimum of wild life that an intensively used area can support: the bittern no longer appears on aristocratic dinner tables since it is rare and has protected status. The drainage of peats means that the soils physically shrink and so water has to be pumped from them continually; this causes them to contract even more and eventually to be so far below sea level as to be very vulnerable to high tides and storm levels, as happened in eastern England in 1953. The organic nature of peat soils and conversely the lack of organic matter in reclaimed heathland soils renders them both liable to strong windblow in spring before they are covered with a young crop. Other wetlands had other fates: the Norfolk Broads, it seems, are largely medieval peat-cuttings for the provision of fuel in a landscape with fewer than normal trees; salt-marshes along the east coast were sometimes converted piecemeal into enclosed salt pans to supply one of the few reliable ways of preserving meat.

One result of all the processes described above was the reduction in the area of woodland. From an early Neolithic ecology in which most of England and Wales was forest and open-canopy woodland, we reach a position by the time of Domesday Book where all the significant woodland could have its size defined, in other words, there were recognized boundaries. The largest single wood was *c.*9600 hectares (about 100 square kilometres), and perhaps 15 per cent of England was wooded, with the highest densities in the south-east and the north. The woodlands were important parts of the resource structure of the communities and nation and so were managed carefully, as described in chapter 3. Though the importance of charcoal as a fuel for industrial processes (especially iron smelting) had a very high profile, the role of woodlands as a source of organic fertilizer for the fields via grazing

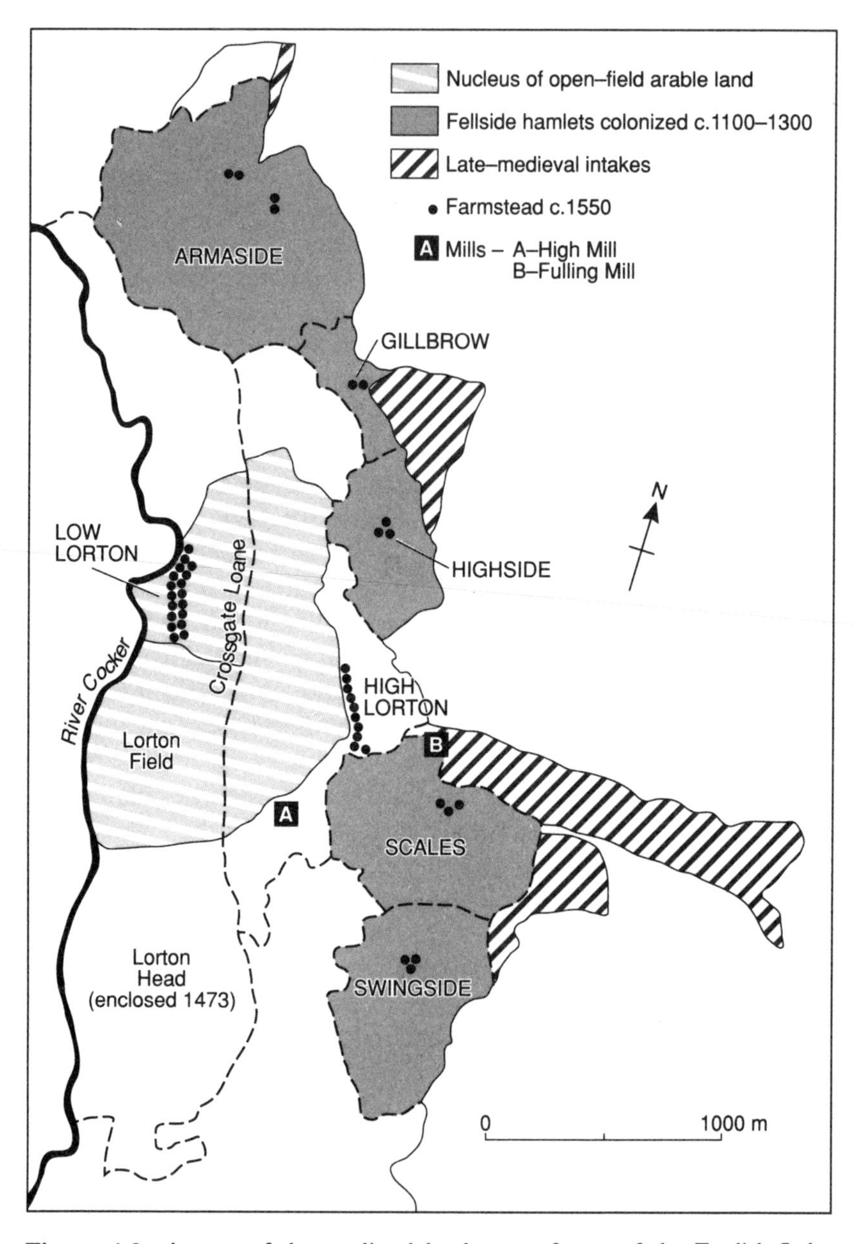

Figure 4.3 A map of the medieval landscape of part of the English Lake District showing the scale and periods of conversion of wild lands to tamed ones. All the shaded area would have been wooded at one time in the Postglacial. (*Source*: A. J. L. Winchester, *Landscape and Society in Medieval Cumbria*, 1987, fig. 34, reproduced by courtesy of John Donald Publishers, Edinburgh.)

domesticates must not be totally buried. It seems likely that as woods were cleared or were sequestered as Royal Forests (see below), this function must have been harder to maintain. Hence the practice, noted as early as Roman times, of 'marling' gained in importance as time progressed. This involves the spreading of soft calcareous material dug from shallow pits; the marl upgrades nutrient status and slows down the inexorable acidification of cultivated soils in a rainy climate. The generally deep soil profile inherited from the precursor forests meant, however, that the desperate-sounding measures of the Mediterranean (such as the dung from the sacred pigeons at Delos) were not needed. Nevertheless, where there were shallow sandy soils not far from the coast, seaweed was a valuable fertilizer. Enclosure of open fields enhanced wood production in one way: as with those areas long enclosed, such as the Celtic west with its thick wall-cum-hedges, shrubs and small trees could be grown in the field boundaries and provide both nutrition and small poles: the hazel was especially valuable. The hedgerow oak might also have enough room to spread its branches into the odd shapes required for ship-building timber, though as John Evelyn remarked in the seventeenth century, it was not easy to keep up this supply. Recourse was therefore often had to imported timber or to placing orders abroad in the apparently inexhaustible timberlands of North America and on the Malabar coast for teak. Admirals who went round pushing acorns into the ground did so in hedgerows and not in deep forests for reasons of ecology as well as personal convenience. Some woods might yield good futtocks, though, because – as Oliver Rackham has shown[6] – a major forest function was to be wood-pasture and the grazing led to an open-canopy woodland in which the bent shapes beloved of the royal shipyards could grow.

Building ships was by no means the only industry in pre-nineteenth-century England and Wales. Iron and its environmental relations has already been discussed and other metals (copper, tin, silver) were important here and there. Lead was an important roofing material and some monastic establishments owned their own veins and constructed smelters. The king owned

[6] O. Rackham, *The History of the Countryside* (J. M. Dent, London and Melbourne, 1976).

Figure 4.4 An area of 'natural beauty' in the Lake District National Park, England. Yet this view of part of Derwentwater contains evidence for management of deciduous woodlands, grazing by domesticated animals, amenity planting by owners of large nineteenth-century mansions, and mining waste. (*Photograph*: I. G. Simmons.)

lead mines on Alston Moor by 1133. The Lake District was home to a number of mines and immigrant delvers from the continent contributed their skills. All such establishments increased the demand for wood, for smelting and also for mine timbers, waggons and the brewing of beer. Other industries with distinctly tentacular environmental reaches included glass-making (a heavy user of fuel), tanning (which needed large quantities of oak bark and animal urine and produced a noxious stream of liquid waste) and textiles, whose mills needed the control of watercourses if they were not to stop work in dry weather. Specialized industries like furniture-making in the Chilterns might turn whole areas of woodland to a favoured species like beech. Given the expansion of industrial production in the eighteenth century, it is not surprising that coal became an important source of energy before the beginning of the century most associated with its dominance as a fuel. Coal-burning had of course contributed to urban air pollution from medieval times onwards, but from about 1300 it began to alter environments in its own right. Even a small bell-pit can produce a landscape not suited to many other uses, and shaft mines with a steam engine, spoil water and nearby housing for the workers are the nucleus of a very distinct transformation of the local environment. Such changes were complemented by increased demand for grassland for meat production and for pasturing the horses that pulled the coal wagons. Further evidence of the control being exerted over nature comes in the eighteenth century with the regularization and canalization of rivers (pioneered by Edward I when he canalized five kilometres of the River Clwyd) so that barges carrying coal (among other goods) could progress quickly and cheaply from supplier to user. The endpoint of this imposition of the human upon the natural comes with the specially constructed canal. The town and the city too are outcomes of the growth of industry and population: a Domesday population of about 1.5 million in England and Wales became perhaps 5 million in Shakespeare's day, 12 million in 1800 and 27 million in 1900. The city in England and Wales, as elsewhere, obliterated much pre-existing ecology and substituted its own, with an enhanced outreach for water (which might be vertical as well as horizontal), food and materials, and an enhanced output of wastes of which untreated sewage and airborne particulates (see Brimblecombe[7] for an account of London) would be the most noticeable.

The environment and pleasure continued to be connected. For the mass of the population, there was no special virtue in rural environments while they were still ubiquitous, though they might provide a certain privacy away from farm and tenement alike for the heroes and heroines of folk-songs, for example, who liked to walk out 'of a May morning'. For the rich, the fencing-off of areas of the countryside from the lower orders was from Norman times onwards an especial pleasure. The Royal Forest, Chases, Warrens, deer parks (35 at Domesday, 3200 in the early fourteenth century), menageries (Henry I had a camel in his), and

[7] P. Brimblecombe, *The Big Smoke* (Methuen, London, 1987).

formal gardens around large houses, were all symbols of a culture as well as preservations of the myth of a Golden Age. The consolidation of large estates by secular landowners followed the example of the pre-dissolution monastic house as well as worldly precursors and led to the position where some landowners were wealthy enough to devote many acres to landscape parks whose production was confined to a few cattle and horses, though the latter might include the all-important speciality of a stud farm. But doubtless there were many country squires like Sir John Middleton in Jane Austen's *Sense and Sensibility*: 'Sir John was a sportsman, Lady Middleton a mother. He hunted and shot, and she humoured her children; and these were their only resources.'

Thus by 1800 the environment of England and Wales was no longer pristine in many parts. If at that date we had been charged with drawing up a list of natural ecosystems (and had been possessed of today's level of ecological knowledge), then it would not have been long. Some sand dunes, some ungrazed salt marshes and mud-flats, cliff faces and unstable river banks would have headed the list. The fens of south Lincolnshire were as yet undrained. The moorlands might have looked wild, but natural they were not. By 1800, natural ecosystems had all but vanished from England and Wales (less so from the Highlands of Scotland perhaps), and the next two centuries were not about to restore them.

3 *The industrial era*

AD 1800 is a useful if arbitrary date: phrases like 'the cradle of the industrial revolution' have a resonance for environmental history. Cradles were much in use: the population of 12 million in 1800 became 46 million by 1950. The mere presence of so many more people brings about environmental change, but of course industrialization meant a greater throughput of materials and energy per capita as well.

The core of the forcing functions of environmental change lies in the city, the conurbation and the factory. These are the seats of innovation and control, the centres of communication networks and the creators of demand. As environment the city itself has not changed qualitatively, but with industrialism got bigger and so affected air, water, land and biota on a larger scale. True, the twentieth century sees the city as a heat store, able to raise temperatures by 1–2°C on a still night; true, it now produces substances unknown in nature such as PAN (peroxyacetyl nitrate), a major constituent of photochemical smog, the scourge of Los Angeles, found in London and Leeds, for example, on those still and sunny days when there are unlikely to be Test Matches. True also that the reach of the ruban area for materials is wide. Whereas the pre-industrial city used relatively small quantities of local materials for vernacular building and imported high quality stone, for instance, only for structures which needed to be imposing, the industrial city could use cheap transport to bring in items such as slates from North Wales to all major cities for roofing working-class housing. At the sites of producition, great mines and quarries gouged into mountain and moorland, and row housing in rural Wales echoed the streets of Wallsend. Specialized regions of brick production such as those using the Oxford Clays south of Peterborough could exist, producing relict wet pits (as those resulting from gravel extraction in almost every river valley) which enhanced the opportunities for wildlife and for the angling which became the most popular working-class outdoor recreation. Brickmaking might also produce a characteristic set of bone deformities in cattle pastured nearby because of the fluorides emitted during the firing.

In the period before 1945, the domestic economy was largely fuelled by coal and so the winning, transport and conversion to power of that fuel was a central economic and ecological feature of the nation. The ecology could be seen in most places, since there was likely to be a coal-fired power station or gasworks even in the genteel towns of Sussex and Kent. The emission control of such places was never very strict for particulates and so downwind residences were both dirty and subject to higher levels of respiratory disease. Coal-winning districts themselves have undergone a number of environmental changes; at first surface mining was the rule, with the formation of bell-pits; then progressively deeper shaft mines, commensurate with the ability to pump out water; and later still a reversion to open-cast mining, this time with huge machines to extract the coal and to replace the overburden. (The getting of ironstone from the East Midlands was akin to this large-scale coal extraction; elsewhere it was mostly very large limestone quarries that came anywhere near the same scale of operation.) Mines produced the sort of local phenomena described in chapter 2; where the rivers were acidified and given extra burdens of silt from pumping and washing processes as well as run-off from waste tips, their effects were further-reaching. Recent open-cast operations are carefully managed to reduce silt loadings, but acid run-off cannot be prevented when water comes into contact with certain geological strata. Many mines had a workforce that lived on the spot, needing housing with water supplies and some sanitation; the coal output was probably transported by rail, adding to land use changes in favour of the habitat of rats and weedy plants. At least one of the latter, the Oxford ragwort, escaped from its site of introduction (from Italy), the Botanic Gardens in that city, by spreading along railway tracks in the latter half of the nineteenth century, perhaps more slowly than other smelly ideas (its Linnean name is *Senecio squalidus*) moving from the same tract in the decade before 1850.

Railways are perhaps the sign *par excellence* of the coming of industrialism and were of surpassing importance in England and Wales. However, apart from consuming land and water, and causing much building, their direct environmental effect is not of great importance, though as a means of fast communication they must have speeded up all sorts of other processes. They were, however, a source of wildfire when passing through heath and moorland areas such as those between Pickering and Whitby or the woodlands of conifers interspersed with the Hampshire heathlands. In the twentieth century the aeroplane and air travel have been resource users. The use of military air power in the Second World War meant that considerable amounts of flat land were taken out of production, thus increasing pressure on the remaining resource.

In spite of a popular perception of creeping urbanization in England and Wales from the 1920s onwards, the proportion of urban-industrial land does not exceed 6 per cent. The environmental significance of such a relatively small proportion has always been (1) in the wastes that have been produced and spread widely, and (2) in the resource demands which the urban population and industrial processes have exerted outside the conurbations and industrial plants themselves.

The same has been the case with agriculture in the 150-year period under consideration. Intensification has been pressed upon the industry by a growing population, even if governments did not really expect that self-sufficiency could be maintained for many foodstuffs; the import of wheat was frequently a hot political potato, as it were. The potato's capacity to achieve more energy per unit area in a cool temperate climate made it very much a star plant of the nineteenth century, though Irish experience in the 1840s showed it to be a fallible crop. One key element in intensification, naturally, has been the application of the concentrated energy of fossil fuels to the land. Steam was never of much value outside the manufacture of agricultural machinery, threshing machines, which did not have to be too mobile, and pumps as for drainage. Oil, on the other hand, powered the more compact and flexible internal combustion engine and so the number of tractors becomes an index of agricultural intensification, along with deeper ploughing and more soil loss. Many other machines were developed which generally compacted soils and increased the rate of run-off, along with the easy delivery of bagged chemicals and the vet. The actual numbers of tractors in the UK took off in the 1930s and peaked about 1955; the line on the graph crosses the downward line for horses in 1950.

Tractors can pull many other machines, and none more successfully than those which apply to the land the various chemical fertilizers and biocides which have become available this century. These make for more stable harvests, both in quantity and timing, essential where industrial processing is concerned. They are cheap, and so large quantities have been applied. Ecologically, excess nitrogen and phosphorus get into surface run-off and groundwater, causing eutrophication in the latter. The post-1945 but pre-1970 generation of biocides, which chemically broke down only very slowly, nearly saw the disappearance of birds of prey like the sparrow hawk and peregrine falcon from England and Wales. Machinery also made small fields uneconomic to till, and so after 1945 many kilometres of hedgerows were grubbed up, along with their flora (often quite diverse) and fauna (birds like the partridge), and with the additional loss of the landscape qualities which seemed to be an eternal rather than a post-sixteenth-century feature of the lowland English scene. Soil blowing in dry weather does definitely increase, however, after their removal. At one stage, the amalgamation of fields and the eradication of small woods was thought to be threatening the survival of foxhunting in the arable shires, but that has been averted, thanks possibly to the influence of the local Conservative Club or even to unexpected visits from men in waxed cotton jackets driving cleanish Range Rovers.

The last two decades of the nineteenth century give us an example of an environmental change which was transient but which must have occurred from time to time in many places in the world: that is, a period of economically depressed agricultural conditions (and climatic recession might have very similar effects) when land previously in intensive use was abandoned to nature and the processes of ecological succession took over. In the English grasslands in the late 1870s and middle 1890s, for example, many pastures became dominated by self-sown coarse grasses and weed species, and even in lowland counties were then fit only as sheep-run or even abandoned altogether for a period. In that case they might soon look like the fields of Essex and Cambridgeshire which were at that time invaded by thorn, briar and bramble, or those of the Breckland, which reverted to the heathland from which they had been reclaimed. Hedges and ditches were likewise neglected, so that the neatness of the landscape was in places lost, though some species of wild plants and animals would have benefited. The structural change which wiped out this phase of reversion was, of course, a shift to a much increased area of permanent grassland in which the demand for milk from the urban areas played a significant part.

Woodlands have not formed a large part of the UK environment in the last 150 years: with Ireland, the UK has consistently been at the bottom of the European league for proportion of land under woodland. With the lifting of the requirement that they supply industrial fuel, many deciduous woods have been unmanaged and a wild wood today is likely to show signs of once having been coppiced (with multiple stems to many trees) along with senescent standard trees. The later years of the nineteenth century saw some small-scale plantings in the UK: perhaps 45 per cent of new plantings between 1884 and 1914 were of conifers, and it looks as if broadleaf forest was not being replaced. In the uplands, the moorland environment was diminished in area by conifer plantations around reservoirs. Demands for industrial uses, especially mine timber, were very strong in the early twentieth century, and then the First World War made heavy inroads into timber stocks. One result was the establishment of the Forestry Commission, which set about replenishing stocks quickly by acquiring cheap land and planting quick-growing conifers, mostly imports from the west coast of North America such as sitka spruce and lodgepole pine, but with Norway spruce as well. Planting policies until quite recently favoured straight lines of trees and ruled edges to blocks, thus producing the kind of opposition whose natural outlet is the letters page of *The Times* on the grounds of landscape changes, interference with access for walkers and diminution of wildlife habitat for birds like the curlew, short-eared owl and merlin. Gradually, some of these areas have had their edges softened and the Commission has successfully attempted to win another kind of clientele by opening up the forests for informal recreation. A number of woodlands have also been managed by local authorities for recreation: the London examples of Burnham Beeches and Epping Forest (transferred to the City of London in 1878) are outstanding. This public use is in contrast to the continued use of private woodlands for sport by landowners and their friends, and increasingly as a business: pheasant shooting is the main example, and here little has changed since the Edwardian days portrayed in Isobel Colgate's novel *The Shooting Party*.

Nearly all facets of industrialization consume more water. The manufacture of a tonne of steel will need 8000–12,000 litres, a car 38,000 litres at the factory, a man's suit 665 litres and a pair of ladies' tights 1.5 litres. Getting this water in England and Wales ought not to be a problem, given the generally reliable rainfall, although the greatest precipitation is generally in the north and west, while the demands are heaviest in the south and east. Supplies in most early industrial phases came from boreholes or rivers, but these were either inadequate in volume or seasonally uncertain. Boreholes and wells, too, tended to produce subsidence as the water was withdrawn, though London's 450 square kilometres depressed by a maximum of 0.35 metres is not in the same league as Tokyo's 2400 square kilometres at up to 4.6

metres. Hence, the dam in the uplands was the preferred solution provided that the strata were not too leaky. Environmentally, such changes drown the pre-existing land use, often farmland, but also exert ecological effects on the river downstream, where there will be less silt, controlled water levels and different water temperature. The water bodies, however, create opportunities for wintering wildfowl and for recreation, though the water companies have always been suspicious of sailors' and picnickers' potential effects on water quality, since potable supplies and industrial cooling water have usually come from the same source. England now possesses the largest such impoundment in Europe at Keilder in Northumberland.

Informal outdoor recreation is generally thought to be a low-impact land use, but this is not always the case: sand dunes easily break down and blow out; moorland and mountain paths act as foci for severe gulley erosion, and heaths and moors are set alight accidentally. Some recreations wear down the habitats: rock climbs get polished smooth and alpine gardeners are reckoned to be major extirpators of rare mountain plants. By contrast, yellowhammer populations often expand dramatically near popular picnic sites in England, filling the niche occupied in North America by the brown bear. So while recreation and conservation are often considered in the same breath as being on the side of Good, they are not always representative of the *status quo.* Conservation as an official activity in England and Wales started in the second half of the nineteenth century with Acts of Parliament to protect sea-birds. (It had been a custom, for example, to hire a boat at Scarborough and fire a blunderbuss into cliffs full of roosting birds.) Although private organizations set up reserves in the interwar period, only after the Second World War was there an official Nature Conservancy with the powers to lease or buy property for the protection of species or habitats. The same Act (1949) also set up National Parks (some seventy-five years after the first such designation, in the USA at Yellowstone) in England and Wales, though not elsewhere in the UK. These parks are in effect protected landscape areas, since they allow a strong measure of development control but at the same time mediate public access to land which is still largely either in private hands or is common land, to which the public has few *de jure* rights of access. The environmental effects (if we compare the parks with similar areas not so designated) are not very strong: smaller quarries and perhaps fewer main roads; fewer large coniferous plantations, even though forestry is not subject to the Act; more clustered housing; more repairs to heavily used footpaths; control of power-boating in the Lake District. But major developments like a nuclear power station in Snowdonia, a large anti-missile radar installation on the North York Moors, together with dams and TV masts everywhere, have not been excluded.

Since no war has been fought on the land surface during this period, the effects of warfare are small compared with Flanders or Vietnam. But preparation for war has been and is important, with large numbers of airfields and military training areas. Live firing excludes recreation; in peaty areas the craters aid the process of breakdown and erosion which is a feature of many upland peat blankets. (The reasons for this general phenomenon are not clear, but grazing densities, acid precipitation and perhaps the physics of wet peat are normally implicated.) On the other hand, the reservation of land from agricultural use has meant that some plants and animals otherwise expelled by ploughing, biocides and fertilizers have survived: many species of Chalk calcicoles survive on Salisbury Plain's tank training areas. One unnoticed effect of the military is under coastal waters, where dumped munitions prevent fishing in several designated zones.

It is not easy to give any concise verdict on the environment of England and Wales during the years from *c.*1800 up to 1956. The region shared with most other industrialized nations the full consequences of that way of life in terms of change due to the use of indigenous resources. Large quantities of wastes have been generated, too, with each individual being responsible for higher totals every decade. On the other hand, as a trading and imperial nation, the local environment could be spared the production of some goods that would have resulted in even greater environmental manipulation: for example, suppose that large quantities of cheap wheat had not been available in the period to 1914. Doubtless, too, the Empire acted as an outlet for other activities such as blood sports, which would otherwise have had to find a target nearer home – on the argument that those who kill wild creatures for pleasure are usually those most in favour of their conservation. The conversion of the land and coastal waters of England and Wales to a humanized environment was just about complete by 1956, but then even in 1800 it had lacked any stretches of truly pristine ecosystems. What has of course happened during this period is that many spreads of already altered landscapes have been changed once again; sometimes the previous use still showed traces in a palimpsest-like fashion, but in others it was obliterated completely. Only when felling takes place does a great forest of spruce reveal that it was laid over an intimate net of field walls; only in the driest summers (such as that of 1984) do the waters of great upland reservoirs dry up to reveal the skeletal outlines of the streets, field walls and building footings of villages like Mardale, lost beneath Hawes Water in the Lake District.

What is distinctive during this period was the articulation of a set of attitudes towards the environment which is not entirely instrumental. We see this in the foundation of national conservation organizations in the great surge of national reconstruction that followed the Second World War. The next, and last, period will trace how they became part of a much more general raising of environmental consciousness.

4 *The 'post-industrial' era*

The key transition took place in 1956–7, when coal (whether used for steam or for electricity generation) was no longer the main fuel of the nation and when indeed a nuclear power station first put electricity into the national grid. Within this last thirty-five years, although the nation is still industrial, a number of important environmental changes have taken place, resulting from the growing importance of oil, the advent of civil nuclear power, a lower tolerance of wastes, a greater dependence on tourism for revenue, and the move towards greater involvement in membership of blocs like the European Community.

Possibly the greatest changes have been in the treatment of wastes, since a great deal of the new environmental awareness has been focused on effluents and much less on the processes and magnitudes of production which spawn them. The general trend has been to improve the dispersal away from the site of production by means of higher, longer and deeper chimneys, pipes and shafts, relying on the general principle of dilution in a high-volume environment like the sea or the air. Thus the gaseous wastes from thermal power stations have been emitted from stacks of considerable height and the sulphur compounds are less likely to fall out on local communities. They are, though, transported up to several hundred kilometres downwind and then rained out. An analogue is found with substances like sewage, which after minimal treatment is piped out to sea via 0.5–1.5 kilometre pipelines, or alternatively taken out to sea by barge as concentrated sludge and dumped. In areas like the North Sea, this has some positive effect, in the sense that the nitrogen and phosphorus enriches the nutrient levels and quite possibly fertilizes the whole food chain leading to commercial species of fish. On the other hand, the sewage contains concentrations of metals which may accumulate to toxic levels in some organisms. Also found are viruses, bacteria and solids which are dangerous to human health inshore. The paradigm case of the argument for dilution is that of radioactivity from the Sellafield (Cumbria) atomic energy installation. Numerous radionuclides are emitted into the Irish Sea by pipeline, all within the limits set by rigorous international agreements. Because of the water movement pattern, much of the radioactivity remains within the basin and is incorporated within bottom sediments. Some then probably finds its way back onshore by pathways which have been difficult to predict, but enough is found in beach sands and estuarine muds to galvanize the authorities into setting lower release limits.

These examples could be multiplied, but are perhaps mainly of interest to Green activists. They do, however, intersect with recent history in the international involvements that they have brought. The Scandinavians, for instance, have been much exercised about the UK contribution to acid precipitation in Sweden and Norway, for example, where many lakes and rivers have been seriously affected. The condition of the North Sea has been the focus of many studies by bordering governments and by the EC: those by the UK government generally conclude that the basin is in good shape, whereas the Dutch find the opposite. The Irish Sea's level of radioactivity is a subject of constant memoranda to Whitehall from Dublin; the Manx government also sometimes takes an independent-minded view on this issue. Thus the question of wastes entails a recognition, as on other fronts, of an imploding world where everything is increasingly connected to everything else. In environmental terms, this response is intensified by the UK membership of the European Community, which is very active on the environmental front. This is not the place to chronicle the recent changes which its directives have brought about, save to say that it is a firm believer in regulatory (as distinct from free-market) approaches to environmental issues, so certain changes in the UK environment over the next few years should be brought about by obeying directives from abroad: the quality of bathing beaches, the amount of nitrogen in freshwater and the provision of environmental impact statements for major projects are all examples.

Since the 1950s, England and Wales have joined in the leisure boom of the industrial world and this has had a number of environmental effects. One is the conversion to mass leisure use of many areas of land: derelict mines become theme parks, country houses with unmanaged woodland become time-share complexes with Scandinavian A-frame houses; large community forests on the German pattern are planned for the fringes of urban areas. Now that some farmland is no longer required for food production, it can be released for recreation, so that farmers have gone into horsey-culture, pick-your-own fruit and clay pigeon shoots, all welcoming onto the farm those members of the public who twenty years ago would have been shown a shotgun in rather short order. The arrival of leisure time and money is highly apparent in the appeal which England and Wales (along with Scotland) make to tourism, both domestic and international. The important historical dimension of this appeal leads to the preservation of past landscapes and buildings, whether in isolation or in a wider setting. There has, too, been something of a vogue for reconstructions of the past in outdoor form, as at industrial museums at Blist's Hill in Ironbridge and Beamish, Co. Durham. Whether these convey anything other than a sanitized and indeed romanticized view of nineteenth-century industrial life is a subject of quite sharp debate: there seems to be a concentration on the coal fire–griddle scones–hooky mat view rather than on tuberculosis and wage-cuts. But all this can be seen in some way as harnessing the environment of England and Wales (in which the rural scene is especially important), in selling most places in terms of history rather than the present – a theme to which we shall turn in the last chapter.

The last thirty-five years have not seen the disappearance of the ways of the previous 150. The contents of the dustbins may now be dominated by plastics and paper rather than coal ash, but the overall contribution of the economy to the global environment remains much the same: whether from oil or coal, the concentration of carbon goes up and promises to change global climate in ways which are probably unpredictable, and if they are, promise no great good to the British Isles. In most of its environmental linkages, England and Wales remain pretty firmly where they were in the nineteenth century: no significant transition to a post-industrial stance has taken place. Economic growth is still the main aim of the government, aimed at producing higher material standards. Environmental matters are still, basically, matters of aesthetics except where world catastrophes threaten; but even there the response is slow enough to raise the suspicion that those in charge of the nation's policy hope that the problem will go away: plans announced in 1991 for linear cities alongside the Thames suggest no great belief in forecasts of a rising sea-level.

Japan: through the looking glass

Comparisons are often made between Japan and the UK. In part this is fuelled by the notion of symmetry, with island nations mirrored at either side of a great land mass. Both are in the Temperate Forest biome, though Japan is much to the south of the UK: in European terms it would stretch from Bordeaux to the Canaries. The analogy is, in fact, closer with North America's east coast: Montreal to Miami: thus the natural vegetation of Hokkaido includes forest dominated by conifers and the coasts of Okinawa include areas of coral reef. Additionally, both are industrial nations with a long and well-known economic history, though in neither case has an explicit national environmental historiography emerged. Here we shall concentrate on the periods of pre-industrial agriculture and of industrial development since 1950 as being pivotal in the emergence of Japan's environmental mosaic.

Figure 4.5 A sand garden in Kyoto, Japan – symbolic of the fact that there are different constructions of nature available in Japanese culture apart from those of the West espoused since the nineteenth century.
(*Photograph*: I. G. Simmons.)

1 *The pre-industrial agricultural phase*

The date of the introduction of agriculture into Japan is not accurately known. The consensus for many years was that rice was the first domesticate and that it came into Japan through northern China and the Korean archipelago *c.* 300 BC. Some recent pollen-analytical work, however, suggests that buckwheat may have been grown even earlier, but it is not yet known how widespread this was. The primacy of rice, however, as the staple carbohydrate of pre-industrial Japan is not in question. Its centrality in Japanese life is symbolized, for example, by the presence of a small patch of it in otherwise formal gardens like the Korakuen Garden in central Tokyo and by ceremonies involving the emperor planting rice. Since wet rice is the usual method of cultivation, the landscape consequences have always been noticeable. At first, only flat riverine lands were converted, but as time went by the fields crept up the slopes and so control of forest vegetation, slope and water flow were all needed. At Toro in Shizuoka prefecture, a mid-Yayoi period (100 BC–AD 100) village has been excavated and shows evidence for elaborate irrigation and drainage systems with sluices, bunds between the fields held in place by wooden slats, and store-houses raised on piles about a metre above the ground. As in many cultures, funerary monuments might also bring about environmental change, and the third to seventh centuries AD saw the construction of large keyhole-shaped tombs (*kofun*): that of the legendary emperor Ninotoku covers 32 hectares in area and is 35 metres high, surrounded by three moats with intervening belts of trees. We can think of the progress of cultivation as a gradual expansion into the forested area. In some places population growth inspired intensification of rice production by the use of draught animals or double cropping; as hardier varieties were bred it became possible to push the limits of rice growing northwards to include the whole of Honshu. One estimate suggests an extension of the cultivated area from 1.49 million hectares in 1598 to 2.94 million hectares in 1730. Such changes enabled the population of Japan to grow from perhaps 6.5 million in AD 1000 to 12 million in 1600 and 31 million in 1730, though followed by a more stable period taking it to 33 million by 1870.

As in so many countries before the advent of fossil fuels, wood was a critical material. Since there was so little external trade during the Tokugawa shogunate (1600–1868), native resources had to suffice for the heavy demands. Apart from the everyday needs of an agricultural society, forests were subject to destruction during phases of civil war and to especially heavy demands in periods such as those after 1570 when *daimyo* (regional governors) began building castles and towns. Such agglomerations often burned down and renewal exacerbated the need for timber: the castle at Edo (today's Tokyo) burned down five times in 270 years. Further, the reverence for *wabi* (cultivated poverty) among the upper castes required high quality materials for the construction and interiors of houses, castles and monasteries. By 1600, woodland throughout the country was harvested to satisfy the requirements of the holders of central power. One consequence in many places was the loss of soils and the flooding of lowland agricultural areas.

After about 1600, there was a change in attitudes and practices. There was an increase in laws and regulations about forest use and attempts by means of sumptuary laws to restrain consumption. A group of forest experts emerged and by the late eighteenth century plantation forestry was a well-known practice, and indeed nearly all forested areas were under some form of regulated use and management system. The land surface was stabilized and forest production was both regulated and maximized.

From the seventh century AD, therefore, until about 1700, the exploitation of Japan's forest resources could be described as 'predatory', as it was by Conrad Totman.[8] In an early stage of the building of large-scale monumental architecture in the Nara-Kyoto area, for example, the watersheds of the Kinai basin were severely depleted, with no interest being shown in the sustaining of yield or of the ecological consequences. One social response was the closure of woodland areas, that is the taking of the mountainous areas into the ownership of monasteries and governments so that some control could be exercised. In the later spans of exploitative forest use, closure was again important, though in the sixteenth century, for example, the *daimyo* used it as a means of ensuring control over woodland resources for military purposes. But scarcity of timber and fuel were not avoided and further attention to the human ecology of forestry was needed by both rulers in cities and peasants in villages. Institutional factors such

[8] C. Totman, *The Green Archipelago. Forestry in preindustrial Japan* (University of California Press, Berkeley, Los Angeles and London, 1989).

as the prohibition on wheeled transport of goods and people between cities may have slowed down forest usage rates and increased transport costs; and during the Edo period, cross-cut saws were banned since they allowed the stealthy logging of reserved trees – an axe, on the other hand, can be heard at a considerable distance.

2 *The industrial period*

Steps to convert the Japanese economy to an industrial footing came after 1881. Initially there was a great growth in the output of the traditional rural sector: in silk, tea and cotton, for example; but after the 1890s basic heavy industry such as iron and steel became part of the scene. The twentieth century has seen the growth of the urban-industrial land use by many times, but Japan still retains about 15 per cent of its land surface in agriculture and 57 per cent in forest: Totman can now call Japan a 'green archipelago'. Environmentally, three consequences of industrialism can be singled out as highly important:

- The postwar strength of the Japanese economy has enabled it to resist calls for the lowering of tariff barriers. Hence it is forbidden to import rice and the price in Japan is perhaps 6–7 times that in the world market. A good deal of agricultural land thus remains in rice which otherwise would be devoted to other crops (which might not require irrigation) or would go out of agricultural use.
- The shortage of flat land was seen as a barrier to successful industrial expansion in the boom postwar years, especially in the construction of iron and steel plants, shipyards and petroleum refineries. Thus in proximity to most cities, land was reclaimed from the sea. All bays and some open coastlines are hence fringed by geometrical slabs of flat land. These have sometimes affected fisheries, since the estuarine shallows that were reclaimed were often nursery grounds for fin fish species or habitat for molluscs. These areas are also vulnerable to *tsunami* (tidal waves caused by sub-oceanic earthquakes) and to severe earthquakes, but have not been very intensively tested yet.

- In the years of the most rapid expansion of GDP, all environmental considerations were forgotten in the scramble to produce. By the early 1970s Japan had a reputation for highly polluted air, freshwaters and inshore seas. This was made worse by certain incidents such as the mercury poisoning of over 100 people at Minamata Bay near Nagasaki, which was first reported in 1956 but not officially recognized until 1962, and the cadmium poisoning of 44 people in the Agano River region of Niigata Prefecture in the late 1960s. After the criticisms levelled at Japan in the course of the UN 1972 Stockholm Conference, Japan addressed itself strenuously to these matters of gross pollution. Ten years later, sulphur oxides in the atmosphere and chemical oxygen demand (COD – a measure of pollution by organic materials such as sewage) had fallen by noticeable amounts. Such changes are partly due to government regulation in response to international and internal criticism and are partly a consequence of differences in industrial structure and in energy conservation following the oil crisis of 1973. In both these cases, regulation seems to have accounted for about half the improvement, but no more. In the case of aluminium, for example, refining was relocated abroad, partly to reduce reliance on imported oil and partly to avoid anti-pollution regulations; hence environmental contamination is exported to another country. However, primary energy consumption had been reduced by 1990 to 80 per cent of its 1970 level.

Overall, the Japanese scene has been cleaned of some of its grossest contaminants, and matters such as noise and smell are often now dominant in the complaints of urban populations.

3 *The Pacific shift and post-industrialism*

In late 1991, Japan became the second-ranking superpower in the world. This apotheosis confirmed two trends with environmental consequences which also marked Japan as emerging into the league of post-industrial economies. The first dates from the 1980s and marks the search not merely for an absence of gross pollution but for a presence of amenity. The latter feature had been notably lacking in cities and their margins in the pursuit of development: urban greenery, waterscapes, roadsides and preservation of historical landscapes were all neglected. In the mid-1970s, for example, London had 30.4 square metres per person of park area, New York 19.2 square metres and Tokyo 1.7 square metres. The leading Japanese city was Fukuoka, with 5 square metres per head. The second characteristic has been the emergence of the Tokyo stock exchange into a leading global financial market and the wealth accumulated thereby. This has enabled Japan to indulge its taste for wood as a form of interior decoration, as plywood in construction and, as a paper-hungry society, for reading and wrapping, at the expense of many of the forests of Southeast Asia and other tropical areas, while maintaining its own status as a 'green archipelago.' Only very recently has there been any awareness of the image which Japan overseas has created in the environmentally conscious around the world by its timber imports and its attitude to whaling.

The awareness of things

Nothing in either Japanese or British culture has prevented either country from making over its land and waters in the pursuit of cultural goals, especially those of material wealth. The western worldview has included a strong strand of humanistic triumphalism, for example, and neither the Taoist nor the Confucian attitudes to conduct seem to have held the Chinese and Japanese back from environmental manipulation when it became necessary or desirable. So both nations stand at the point where they can still talk of 'balances' and 'trade-offs' between economy and ecology: the thought of limits to the former caused by the latter are not within the conventional and governmental patterns of thought.

This last, perhaps rather strongly worded, assertion makes the point that the question of the relations between human societies and their environments is complex. So far, this book has largely treated the question as linear in the sense that the natural sciences provide information about the natural world and the human manipulations of it. People (individually or in various groupings) then respond to these data, either by believing the science and acting in accordance with its predictions (a 'rational response') or by doing something else, like ignoring it as too uncertain ('politically motivated') or as not fitting in with a western cosmology or world view, which is usually labelled 'traditional' or 'alternative' behaviour. Thus, the cultural framework into which environment has fitted throughout the histories chronicled empirically here is important not only as background for 'objective' accounts but as a precursor to present-day interactions. In the final chapter, we shall examine some of the cultural attitudes to environment that have been present during the last 1000-odd years, mostly in the West. The account will look firstly at the notion of wilderness in its purest state, as nature completely untrammelled by humans; then at the wild as a desirable ingredient in the culture of a civilization; and finally at the cultural evaluation of nature in general.

5

'*Our good landscapes be but lies*': Culture, Time and Environment

Although the scientific-technological view of the natural and human worlds is dominant at present, we need to recognize that this is a relatively recent phenomenon. Indeed, it has only reached a full flowering in the last 150 years, despite its earlier origins. There are counter-cultural groups who do not subscribe to it and there are societies which are unaffected by it intellectually, but it is unlikely that there is anybody in the world who is not in some way touched by some facet of its works. There is a long history of comment on the natural world and its values. We need to examine this in the light of the fact that what people say about nature is not necessarily how they behave towards it. Chinese Taoism, for example, as seen in the *Tao Te Ching* of the sixth century BC, envisages a quietistic, non-interventionist role in the natural world, but that did not prevent the Chinese of the time from making enormous changes to the land and water around them.

The basic thread of our story here has been the transformation of natural ecosystems into humanized ones. This chapter will deal firstly with the wildest of all – the areas apparently unaffected by human activity, which are today usually labelled as wilderness.[1]

[1] G. H. Stankey, 'Beyond the campfire's light: historical roots of the wilderness concept', *Natural Resources Journal*, 29 (1989), pp. 9–24; R. Nash, *Wilderness and the American Mind* (3rd edn, Yale University Press, New Haven, CT, and London, 1982); M. Oelschlager, *The Idea of Wilderness: from prehistory to the age of ecology* (Yale University Press, New Haven, CT, and London, 1991).

Figure 5.1 An aristocratic hunting park in Weichang, China, until the area was given over to peasant agriculture in the nineteenth century and disforested. Soil loss and deep gulleying of the slopes, aggradation of the river channel and general economic uselessness apart from grazing are now apparent.
(*Photograph*: Professor Pu Hanxin.)

This leads to the question of cultural views of wild lands as an element of global land use and of social patterns, and this in turn merges with a consideration of the whole constellation of relationships, past and present, between humans and their environments. An issue here is whether we are right to go on accepting the division implied by 'and' in the last phrase of the previous sentence, or whether a more appropriate word might not be 'in'.

Wilderness ways

Some readers will recognize this phrase as the name of a chain of shops in the UK which sells clothing and equipment for outdoor

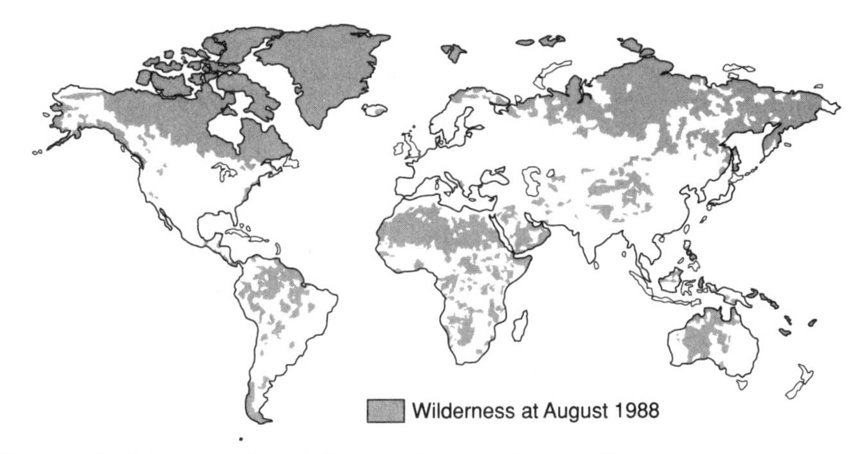

Figure 5.2 The remaining wilderness: the world in the 1980s – a reconnaissance survey.
(*Source*: M. McCloskey and H. Spalding, 'A reconnaissance-level inventory of the amount of wilderness remaining in the world', *Ambio*, 18 (1989), p. 223.)

recreation. No piece of land untouched by humans exists in those islands, yet the word has a resonance which appears to attract the customers. In North America too, the word 'wilderness' has an especial appeal. Perhaps this derives from the fact that where (as in the USA and Canada) it is still possible to find virtually pristine landscapes, these have acquired a distinctive value. Let us then explore the geography of present-day wilderness and then look at the cultural background that has brought wilderness to the level of attention it currently demands.

Figure 5.2 is called by its compilers[2] a 'reconnaissance-level' map of the amount of wilderness remaining in the world. They define wilderness in a way which follows a piece of United States legislation, as land that 'generally appears to have been affected primarily by the forces of nature, with the imprint of man's work substantially unnoticeable'. The basic methodology of the map was to take a series of US Defense Mapping Agency navigational charts at scales of 1:1 million and 1:2 million and eliminate all areas which showed evidence of human construction, together with tracts where agriculture and forestry, past or present, could

[2] M. McCloskey and H. Spalding, 'A reconnaissance-level inventory of the amount of wilderness remaining in the world', *Ambio*, 18 (1989), pp. 221–7.

be inferred. Only large blocks of land, of at least 400,000 hectares (1 million acres) were selected. An element of approximation is thus introduced which points us to a general background consideration: that of scale. The world map is inaccurate because it omits smaller units, but whether by a factor of 10 or 20 (or more) we do not know. There is of course a further element of inaccuracy since the land use and land cover patterns of the earth are not static. Since the compilation of any map, there is likely to have been colonization of some areas for mineral extraction or the erection of a radar installation, for example, or even the abandonment of such structures and their reclamation by the processes of the non-human world. We can also see that figure 5.2 leaves the oceans out of consideration altogether.

Conceptions of wilderness

The definition of wilderness used in the US Wilderness Act of 1964, enshrines the essential qualities of the word and its use in English. It is essentially a place where human influence is limited, either because it has never been present, or because humans have come and gone. It is land without permanent structures or roads, is not regularly cultivated nor heavily and continuously grazed, but may have been subject to seasonal pastoralism. In Australia, a gradient of 'biophysical naturalness' can be used to identify at the one end areas unlogged and ungrazed which are remote from settlement and access routes and free from permanent artifical structures and so qualify for consideration as wilderness.[3]

The origin of the word 'wilderness' is very likely *wil(d)-déorness*, which means 'the place of the wild deer' in the Old English version of the word. (This is not to say that wild ungulates are found in every ecosystem shaded black on figure 5.2, but outside the ice of Greenland they can be seen in most of them, Australia obviously excepted. However, the Oxford English Dictionary allows that wild deer 'might in medieval times mean simply wild animal'.) In this version and in its Germanic predecessors such as

[3] R. G. Lesslie et al., 'A computer-based method of wilderness evaluation', *Environmental Conservation*, 15 (1988), pp. 225–32.

Wildern, the meaning approximates to our current definitions of savage and uninhabited places left to nature, though not without vegetation, for which the term desert was favoured. In seventeenth-century England, though, it was used of a maze or labyrinth planted with trees or hedges in a park – an elaborate reconstruction, we might think, of what had in that country largely vanished. This range of meanings can be echoed in most modern European languages and their Classical forebears, and so it is of interest to look briefly at other languages which have spawned a literate high culture to see if they extend the concepts.

In Sanskrit, a language which has a common ancestor with many Indo-European tongues, the equivalent words are:

Araṇya	A wilderness or desert equally: the wilderness could be of forest.
Vana	A forest or grove, which sometimes refers to a thicket of e.g. lotus or other plants growing in a thick cluster.
Jaṅglabhūmi	A land covered with thick jungle.

None of these appears to have any particular value-connotations: they are simply descriptive labels of particular types of terrain.

In Arabic, unsurprisingly, the words are often related to the conditions of the desert biome with its open character and horizons. Words of relevance include:

Bādiyah	Related to the word Bedouin, with a notion of openness.
Bariyah	A cluster of meanings around the concepts of earth, dust, smoothness and exhaustion.
Falāh	A waterless plain, a journey or search, an open space.
Mafāzah	The desert as an escape or place of refuge.
Sahrā'	A tawny-coloured desert or broad land.
Baidā'	A dangerous desert or perilous place.
Tayhā'	A trackless wasteland in which to wander astray.

Here we seem to have two clusters of words: those which describe the desert or steppe terrains of the heartland of the language, and those which venture the rather negative side of such places in which there are few guide-posts, few tracks and in which it is possible to get lost. However, with the desert as a place of refuge, there is some sign that there are positive values. This again is not surprising, given the adaptation of the pastoral economy of many other groups to the deserts of the Middle East and Africa.

In Chinese, the word for 'wilderness' means the place where people have never or rarely been and which is so far kept in the natural state. It may thus be associated with an emptiness such as desert or the haunt of beasts. Transferred meanings may include remote, rural or sparsely populated areas: such would be the meaning used by a city or town dweller. In a more restricted sense, the characters for 'fieldwork' imply a survey in wilderness areas, even though the term can be used in an urban context. In a wider sense, the qualities of neither paradise nor fearsome place are imputed to wilderness: it is a simple, matter-of-fact description. In Japanese, there are a number of words which relate to the concept of wilderness, of which *arano* is the closest. The *ara* part of the word (*no* means 'field' or 'rural place') may mean 'harsh' or 'rough', but it also has an etymological connection to the verb *aru*, 'to be born, appear, grow'. So although *arano* means 'waste land', 'uncultivable land', there is a subtextual implication of something being born or growing there.

What so many of these etymologies and usages have in common, especially in western languages, is the implication of opposites, like the desert and the sown, the wild and the tame, a wilderness and a meadow. In the wilderness the non-human world is undefiled and even paramount. This can itself lead to another polarity, that of human attitudes towards an unaltered state of nature.

Wilderness as desert

Much of the western view of wilderness is evoked by Edmund Burke in his *Speech on Conciliation with America* (1775):

> By adverting to the dignity of this high calling, our ancestors have turned a savage wilderness into a glorious empire . . . by promoting the wealth, the number, the happiness of the human race.

Such an attitude derives, in the Christian West, from many centuries of regarding the Creation as unfinished. The job of humanity was to make the rough places plain. For example, Judaeo-Christian sources are unambiguous in their views of wilderness. The Book of Deuteronomy (32:10) talks of 'the waste howling wilderness', and St Matthew (11:7) makes an implicit contrast between nature and eschatology:

> what went ye into the wilderness to see? A reed shaken with the wind? . . . A prophet? Yea, I say unto you, and more than a prophet . . .

– a view which seems to be confirmed by the location of Christ's temptation in the desert. It is not surprising that these ideas carried forward into the influential writings of the desert fathers. These were Coptic monks of the fourth century AD who inhabited the solitary places of Egypt and Palestine and chose to live an ascetic life near to the limits of human endurance but close also to animals, angels and demons. The desert, for them, was to be the place of the final warfare against the devil, especially as presented in the form of their own desires and wishes. Hence a life of radical simplicity and integrity was essential, since the monk was to be separated from all and yet united to all. The *Sayings* of the Desert Fathers show that they held the desert and its animal inhabitants in high regard, but it is clear by implication that a sparse ecology was conducive to an ascetic life-style: it was austere and silent, as were the monks.[4] Above all, there were fewer people and fewer traces of human habitation, a function of wilderness which is alive in some of today's evaluations. A reassertion of the desert values into the more lively world of medieval Umbria can perhaps be seen as the environmental piety of St Francis of Assisi.

[4] S. P. Bratton, 'The original desert solitaire: early Christian monasticism and wilderness', *Environmental Ethics*, 10 (1988), pp. 31–53.

The year 480 saw the birth of Benedict of Nursia, the founder of the Benedictine Order, which also carried on some of the desert traditions in founding its monastic establishments in wild places away from the potential corruptions of normal human existence. This was particularly so during the Cistercian revival which planted abbeys in forests, moorlands and wetlands all over western and central Europe after the end of the twelfth century. However, if the cities were especial abodes of sin, these rural retreats were not beyond improvement: Clarence Glacken[5] reproduces for us an account of an abbot and the monks going out to fell trees, with each group of workers having an appropriate Latin name and the whole enterprise being sprinkled with holy water and incense.

An Asian culture like that of Japan contains examples of similar attitudes. In Japan, Kamo no Chomei (1153–1216) abandoned the world for a simple hut on Mount Toyama, where in summer he might hear the cuckoo call, 'promising to guide me on the road to death'. In the meantime, however, the natural history of the mountain gives pleasure, for 'my only desire for this life is to see the beauties of the seasons'; yet even this attachment might be contrary to Buddhist teaching and hence a hindrance to salvation. In the Tokugawa period, the famous Japanese *haiku* poets such as Basho (1644–94) similarly extolled the wild:

Haranaka ya	On the moor, from things
Mono ni mo tsukazu	Detached completely –
Naku hibari	How the skylark sings!

but were happy when the smaller-scale and mundane transformations were encountered as well:

Kuromite takaki	Sombre and tall
Kashi no ki no mori	The forest of oaks
Saku hana ni	In and out
Chiisaki mon wo	Through the little gate
Detsu iritsu	To the cherry blossoms

[5] C. L. Glacken, *Traces on the Rhodian Shore* (University of California Press, Berkeley and Los Angeles, 1967).

If the monastic establishments of the West embraced their perceived wildernesses as containing fewer worldly temptations, the common people were apt to regard such places as distinctly undesirable; the forests and the wetlands, the large lakes and the wild sea were all potential sources of terror as well as of produce that might be hunted, gathered or pastured. In all, an adversarial relationship seems to have prevailed. A good example is the eighth-century Old English poem *Beowulf*, which deals with life and heroism in an Anglo-Saxon society. In it, the depths of the forest contain a large lake, home of the dreaded monster Grendel and his mother; the uplands were the haunt of the dragon which was to kill Beowulf after ravaging the country. Grendel clearly lived in a waste place:

> Now Grendel, with the wrath of God on his back, came out of the moors and the mist-ridden fells with the intention of trapping some man in Heorot.

When tracked down, Grendel's mother's lair was reached over broken, rocky ground, over narrow, forbidding bridle-paths, beetling crags and, eventually,

> Suddenly he came upon a dismal grove of mountain trees overhanging a grey rock. Below lay the troubled, bloodstained water of a lake.

And the dragon, along with feuding, was responsible for 'the empty banqueting-hall which is now cheerless and the home of winds. . . . The harp makes no sound and in the courts there is no longer joy.'[6]

These concepts echo Old English elegies like *The Seafarer*, whose life was blighted,

> Lacking dear friends, hung round by icicles,
> While hail flew past in showers. There heard I nothing
> but the resounding sea, the ice-cold waves.[7]

[6] *Beowulf*, trans. D. Wright (Penguin, Harmondsworth, 1957).

[7] R. Hamer (trans. and ed.), *A Choice of Anglo-Saxon Verse* (Faber and Faber, London and Boston, 1970).

or the protagonist of *The Wanderer*, who ponders

> When all the wealth of earth stands desolate,
> As now in various parts throughout the world
> Stand wind-blown halls, frost-covered, ruined buildings.[8]

For these unknown writers, the wild stands for decay and destruction: it is the opposite of fruiting and growth. It can be part of the cosmic scheme of things, but may also be the result of the wrong kind of human activity as well. In the twelfth-century *Romance of Tristan and Iseult*, the illicit lovers can find a home together only in a forest wilderness: instead of living amidst silks, they sleep among brambles. The wilderness is thus a world which is the opposite of a properly ordered human society and a punishment for those who break the rules.

Nowhere, possibly, was the duality of civilization and wilderness more marked than in the Puritan settlements of New England in the seventeenth century. In mimesis of the Exodus, wilderness tested the faith of the settlers in God and represented a defiance of Satan. Fruitfulness was to be obtained by the 'cleansing away' of the woods and their replacement with 'goodly meadow'. The clearing of wilderness might include the aboriginal inhbitants, since they had neither enclosed land nor domesticated cattle. Carolyn Merchant[9] argues that this type of perception flourished as nowhere else in New England, not least because as prosperity followed trade in fish, fur and lumber there was a need for a symbolic and rhetorical wilderness to recall people to their former spiritual condition.

Wilderness as a paradise

If we look to the opposite side of the valuations, no more popular an evocation of wilderness as a delightful escape from all the pressures of worldly existence can be found than in the Fitzgerald translation (1859) of *The Rubaiyat of Omar Khayyam*:

[8] Ibid.

[9] C. Merchant, *Ecological Revolutions. Nature, gender and science in New England* (University of North Carolina Press, Chapel Hill, NC, and London, 1989).

> . . . and Thou
> Beside me singing in the Wilderness –
> And Wilderness is Paradise enow.

– which appears to pick up on William Cowper's imagined eighteenth-century retreat, in which he might find

> . . . a lodge in some vast wilderness
> Some boundless contiguity of shade
> Where rumour of oppression and deceit
> Of unsuccessful or successful war
> Might never reach me more!

and is again echoed in Gerard Manley Hopkins's *Inversnaid*:

> Long live the weeds and the wilderness yet.

These well-known lines are the heirs to a tradition of praising nature which dates back to, for instance, Plato's *Timaeus*. Here, nature brings order into the universe to form a true cosmos. It is nature which assures the proper functioning of all the components of the world, and to follow nature is to ensure plenty. This whole view point was elevated by the action of Petrarch in climbing Mount Ventoux in Provence in 1336 in order to gain a sense of stability amidst the troubles of his life. Of course, praising nature did not always mean the natural world unaltered, but at the root of exalting the country estate and farm there was still the notion that there would be found inherent qualities of immense value to mankind.

From such a set of roots has grown the basis of the modern western attitude to wilderness. This is the view that underlies the map with which we started: not the replacement of wilderness with 'useful' and 'productive' land, but its preservation in order to retain the cultural values inherent within it and to contribute to some perceived ecological equilibrium on a global scale. The cultural and ethical values of wilderness are exemplified by many of the writers quoted above and were brought to the fore particularly in the campaign for a statutory Wilderness Act in the USA in the early and mid-1960s. These non-utilitarian values can

Table 5.1 Reasons for wilderness preservation

UTILITARIAN VALUES
Scientific
Preserving a sample of ecosystems to ensure biotic diversity
Conserving gene pools and potentially useful organisms
Protecting natural areas for research and monitoring
Economic
Providing a particular form of recreation
Conserving wildlife
Protecting watersheds and water quality
Conserving scenery for tourism
Avoiding diseconomies of development
Promoting a 'balanced' land use pattern
NON-UTILITARIAN VALUES
Cultural
Conserving a cultural heritage
Preserving aesthetic values
Providing educational opportunities
Ethical
Rights of non-human entities
Scope for individuality
Social value of exercising restraint over change and transformation

Source: Adapted from L. C. Irland, *Wilderness Economics and Policy*, Lexington Books, Lexington, MA and Toronto, 1979, p. 2.

be summarized as being a number of types (table 5.1). The first is usually described as spiritual. Here is the chance to be at one with untrammelled nature, an experience which may be described as religious by those who wish to use such language. The nature of this religion may not be conventional: for example, it was used in the 1960s by pot-smokers in Yosemite National Park who had come to rap with Mother Nature in an elevated condition (theirs, not hers). But time and again the idea surfaces of being close to a god or its equivalent in places which are as wild as possible and thus removed from human artificiality.

The second is more conventionally recreational. Enthusiasts like to think there is a special quality to their journey (on foot or by raft, typically) if it is undertaken in a small party of people who encounter few others on the way. Thus in some managed wildernesses, entry is rationed and groups are spaced out along the trails so that they are out of sight of one another, in a manner reminiscent of the intentions (if not the visible practice) of air traffic control. These buffs are usually the most willing to encounter an unmediated nature: 'take nothing but photographs, leave nothing but your footprints' is their physiologically challenging motto. For them the hazards of an unexpected bout of appendicitis or a forest wildfire are part of the acceptable risks of wilderness travel. They do not expect an emergency telephone every few kilometres or to be lifted out by helicopter, except in such extreme circumstances that the managing authorities would fear being expensively sued by surviving relatives if nature were allowed to take her course. This set of attitudes has not, however, prevented the launch in 1990 of a *Journal of Wilderness Medicine*, including papers such as 'Normal oral flora in black bears: guidelines for antimicrobial prophylaxis following bear attacks'.

The paradisaical side of wilderness preservation is also seen in some more utilitarian views, where the needs of scientific research and the possibilities of new biotic materials (such as genetic diversity or plants with useful properties) form two ends of a spectrum of more instrumental attitudes. Thus we are told of the need for pristine ecosystems as datum-lines against which to measure human impact. Similarly, we are informed of the requirement to learn how best to manage natural resources by studying the functioning of 'natural' ecological systems. This view embraces the arguments for protecting, for example, 'virgin' rainforests which include estimates of the number of plant species which may contain useful substances such as pharmaceuticals. It is estimated that if we take a prescription to a drugstore, then there is a 25 per cent chance that the original active component had its origin in the plant from a tropical forest.

So we have here the complementarities of a paradise: that it should be beautiful and also useful. Such was the original sense of the word, since it implied a type of garden; however, in the original Persian form the garden was walled, which would never do for today's wilderness enthusiasts, whether recreational or scientific. The balance of these two views will be examined again at the end of this chapter.

The main currents of thinking about the why and how of wilderness have for about 100 years come from the USA. The pace of environmental change in that nation, the rapidity with which the aboriginal cultures were replaced with a new set of world-views, and the reaction to those events in the form of historical theorists such as Frederick Jackson Turner (1861–1932), poet-philosophers of the wild such as Henry David Thoreau (1817–1862), and philosopher-activists such as John Muir (1838–1914) have placed the USA firmly in the avant-garde of thinking about wilderness, and also of designating and managing areas of 'wilderness'. It is the North American thinkers who have been instrumental in changing our thoughts on the subject in the second half of the twentieth century. They inherited, of course, a strong religious tradition in which the wilderness was a place of purification away from the strife and temptations of the settled areas and especially from the city. The advances of science and technology had made the wilderness less of a fearful place: as it was better understood (and more capable if necessary of being controlled), then a more benevolent attitude could be adopted towards it. Then there is the paradox that only when wilderness is scarce is it valued as a resource: while natural ecosystems are the norm, there is no special value accorded to them. Finally, lest it be thought that the thinking has dissolved into a kind of cosy 'be-nice-to-wild-things' view, there is a position which regards wilderness areas as prisons, as incarcerating wilderness. Here, it is argued,[10] the designation of wilderness acts as a control over wildness, keeping it in carefully subdued areas which do not threatent the rest of the imperium of human hegemony over the natural world: by keeping a fenced-in wilderness, it can be enjoyed because it will not escape to threaten the rest of the human domain. An analogy might be with the tigers in a modern safari park.

[10] T. H. Birch, 'The incarceration of wilderness: wilderness areas as prisons', *Environmental Ethics*, 12 (1990), pp. 3–26.

The call of the wild

Less demanding than a call for the presence of wilderness is the set of cultural attitudes which reckons that wildness is a necessary part of the human environment. There is no need for the ecosystems to be pristine, but they need to be perceived as being dominantly natural. This may lead to an interesting role for studies of environmental history in pointing out the discrepancies between valuations of regions as wild and the fact that they may be some degrees removed from being truly natural ecosystems.

The historical role of the wild in human psychology has been subject to a number of imaginative reconstructions. At one extreme is the group which places great emphasis on human–environment interactions during the hunter-gatherer phase of economic existence, which accounts for 90 per cent of our evolutionary time as a species. It can be argued that the qualities needed for success in that way of life are such things as immediate response to short-term events and a premium on aggressive behaviour among males. These would have militated for the greatest success in hunting, for example, since rapid response to the presence of game and rapid reactions in actually killing it would have been essential. Likewise, working up enough aggressive emotion to pursue the chase to a sucessful conclusion would have been a useful trait, even if this mainly involved digging large holes for pitfall traps. The persistence of these psychological attributes into the twentieth century has led two commentators[11] to postulate the present-day actuality of a 'boiled frog syndrome' in which the human species is unable to respond to the true nature of today's problems, which are long-term and not immediately perceived, unlike those of the hunter. In this view humanity is like the frog placed in a pan of cold water which is then heated up gradually: it doesn't recognize that the time has come to jump out until it is too late. The aggression theorists tend to the view that young males especially must have some way of discharging the innate aggression which they have inherited and that to lack those opportunities

[11] R. Ornstein and P. R. Ehrlich, *New World New Mind. Changing the way we think to save our future* (Methuen, London, 1989).

will inevitably bring out the aggression in a less sociably acceptable form, such as trashing the local trees or football hooliganism. The cure tends to be more games or wearying slogs up mountains with heavy packs, thus making for large numbers of young people later determined to concrete over everything in sight.

A variation on this reconstruction of history is provided by Paul Shepard in his thesis that the environmental history of humanity is intimately linked to the development of the child. A key element in this is access to nature in the pre-adolescent years, so that the child is completely at home with the 'other' in the form of the wild (including imaginary identification with animals in the form of games), together perhaps with domestic animals. Failure to have this set of experiences allows for a rejection of nature in later life and ends up in the dominance hierarchy with nature at the bottom which is theorized by some commentators out of the empirical history of the west. Not all writers on such themes agree with overarching reconstructions of 'psycho-history', pointing to the great variety of philosophical views and ethical attitudes towards the environment that have been characteristic of most literate cultures since Classical times in both East and West. This suggests a plasticity of behaviour in which the inheritance of the hunter is not so dominant, in which there are in effect choices about how to act with regard to nature. This squares well with experience, except possibly for the caveat that what writers write is not necessarily what the great majority of others actually do. The letters page of *The Times*[12] has hosted numerous complaints about coniferous forests in upland Britain, for instance, but a large number of people unreasonably persist in enjoying their visits to them.

Most literate cultures have at some point had a school or individual who extolled the virtues of the wild; it would be tedious to try to list them all. As with wilderness, there is the general point to be made that while wildness is ubiquitous it is not worth mention by any kind of elite, but when it starts to become remote from major settlements then a revaluation takes place. Part of the valuation of the wild is no doubt strictly practical: there are certain activities which need less intensively-used terrain for their

[12] Of London, that is.

success. In medieval Europe, this might be hunting animals which required forest, or at any rate little influence from settlement. A present-day equivalent might be a noisy sport like motor-cross which in both its decibel level and its effect on terrain requires land which can be spared from production. Walking for pleasure can be carried on in the interstices of a productive landscape, but requires for greatest appeal the presence of something challenging to be achieved, and that preferably it should have at least a hint of the wild about it: Petrarch's scaling of Mount Ventoux in Provence had much in common at that level with the intrepid British who, clad in tweeds and ropes, went up most of the major Alpine peaks during the second half of the nineteenth century. So the valuation of the wild through history will contain the very practical element of free-market economics, in which its value rises as it gets scarcer, and that scarcity will vary, among other things, with the friction of distance. That in turn will get lower as transport gets faster and cheaper; the effect of the railway upon the valuation of wild country probably exceeds by several orders of magnitude the sort of actual environmental impact alluded to in the previous chapter.

All this suggests to us that the channels through which societies receive their messages about the environment must be important. Direct experience may be lacking, or the meaning of any experience may be mediated through cultural channels: we are told that to respond positively to a particular scene is the proper thing to do. For some societies, oral tradition, graphic arts and myth may have dominated this added layer of meaning; to this was added first writing and printing and now the full panoply of electronic communication, of which television is probably the most important in forming our view of 'the environment'. (All those falling trees! All those coral reefs! All those men [*sic*] in safari jackets!) A portion of the cultural ecology of the wild, therefore, is its representation in terms of aesthetic appeal. We saw a hint of this in the discussion of wilderness-oriented literature, but it cannot be any revelation that the field is wider: all forms of aesthetic expression have at some time dealt with what was generally termed 'nature', and included in that the relationship between the wild and the tame, between 'the environment' (as we would now call it) and humans. Kenneth Clarke said of landscape painting, for example, that the tacit question is always 'what is man's relationship to nature?'[13] Such commentaries find it difficult to disentangle high culture from economic considerations, seeing them as all part of a single whole. Art, religion, myth, occupying the day, eating and sleeping are not necessarily separated from each other, so historical discussions tend to be indeterminate at the edges, rather like a forest clearing after it has been abandoned.

In his account, Paul Shepard[14] roughly parallels the periodization used in the present work. He notes that for hunters the landscape is full of sacred places which are intimately tied to the myths of the culture. Myth in turn confers orientation both on topography and on concepts, and language confers words upon units of experience and on places. Language, then, is the foundation of both beauty and utility: it is not an inventory but a creative symbolic organization which imposes meaning and orientation. Neither myth nor language are, of course, absent from later stages of human cultural formation and so perhaps some continuities may be preserved.

Pastoralism of the nomadic kind brings about long periods of leisure (for the men at any rate) interspersed with all-consuming activity. There is time therefore to develop art and hedonism and conversation which is abstracting in its tenor. It can be directed towards an etherealized art and science in which the mythos is a landscape dominated by the sky which is the promise of order and orientation. It probably represents something of a Golden Age to the fixed farmer, who has so little leisure and who is also supporting the people of the city. Thus from Classical antiquity onwards there has existed a yearning for the pastoral. The apogee of this life-style is perhaps the image of the warm Mediterranean afternoon in which there is time to discourse, to think, to make music and love. In medieval times, it was the bucolic answer to the cold steel of chivalry; in the seventeenth century it promised a deeper perception of nature, as in the poetry of Sir Philip Sidney, just as in this century a sharper but contiguous view is found in W. H.

[13] K. Clarke, *Landscape into Art* (John Murray, London, 1949).

[14] P. Shepard, *Man in the Landscape. A historic view of the esthetics of nature* (Knopf, New York, 1976).

Auden. The Mediterranean afternoon mythology might also be a major element in the best-selling success in the 1990s of books about English couples settling down in a Provence that sounds suspiciously like a re-creation of *Daphnis and Chloe*.

The agricultural phase of human existence also gave us the garden, which is another enduring source of aesthetics. One of its great virtues is its combination of beauty and utility, mentioned above; another its appeal as a contrast – of a watered oasis in dry lands, or a regularized space among the apparent orderlessness of the forest. Its apogee in the pattern of the formal garden seems to come as a symbol of control over nature, but it is possibly more complex than that, for such gardens were regarded in their time as 'natural'. This may stem from the perception of the garden as a perfection of nature with human help, and it applied to the less formal English landscape park as well. According to some interpreters, the garden's appeal is that of the female both as mother and sexual object.

Travel brought different notions of what was aesthetic about nature: the landscape as an object in itself seems to have been a Dutch invention of the seventeenth century, exported in painted form. In fact, at that time humans tended to withdraw from the picture and turned to look at it, in an interesting parallel with the development of the epistemology of the natural sciences. The picturesque brought together music, literature and gardens in forms in which nature was held in highest regard when it resembled cultural notions of high art.

Industrialization based upon science (in its sense of *scire*, 'to know') and technology sets up a number of different aesthetics to represent high value. Whereas medieval nature might be a set of pictures and Renaissance nature modelled in terms of mankind ('humanism'), later writers saw it as a machine with inevitably largely male characteristics. This reached its apogee in the kind of land survey which was applied to North America, in which abstractions are used to organize perception (cf. the development of theory in science) and the land is experienced as a series of straight lines, where anything that does not fit the pattern can be removed; literally so once the bulldozer had arrived. The process of abstraction then permits its users to regard 'resource' as something separate from the rest of nature, to be used ('wisely' no doubt)

Figure 5.3 Claude Monet (detail), *Argenteuil*, *c.*1872, canvas, 50.4 × 65.2 cm. Monet painted the onset of industrialism in France during the 1870s and 1880s, but thereafter turned his back on such developments and concentrated on an increasingly abstract view of his water-garden. In this painting, the old-order chateau is neatly sandwiched between two chimneys of new-order industrial plants.
(*Source*: National Gallery of Art, Washington, Ailsa Mellon Bruce Collection.)

whenever economics allow, but without regard for the rest of the web of which it had been a part. A parallel lies in the rigid constellation-like movements in the music of Haydn and Mozart and then the pared-down abstraction of the serialism of Schoenberg and Webern, followed by the minimalists of the second half of the twentieth century (Cage and Birtwistle, for instance), who provide the musical equivalent of the Slough trading estate. In philosophy, this trend towards the separation of humanity and nature appears in the form of existentialism, which can be seen as a reaction to generalization and abstraction, to the finding of norms and averages and the reduction of both nature and humans to statistical traits. Philosophers can then argue that existentialism becomes an assertion of solipsism where only the 'I' is real and it can assert a bold individualism in the face of almost anything else. The end-point, though, is a firm separation of the inside and the outside: there are the humans and there is the environment, and little importance is accorded to the 'outside' within in the shape of the chromosomes or its effect on blood and brain.

These notions of what appeals to the emotions, coupled with the more systematic view of knowledge which we call reason, come together in a series of constructions of the notion of environment which societies make. We shall conclude by trying to see which of these constructions have been in play during human history, often as part of the unquestioned outlook of the societies of the time.

Attitudes to nature through time

Somehow or other human societies construct the environment for themselves. To do this in a simple, pre-industrial society involved the use of material that was largely mythical and orally transmitted. By contrast, one of today's industrial communities will receive a big input from the latest findings of the natural sciences. We cannot communicate directly with very much of what lies beyond our skins, of course: unlike Dr Doolittle, we cannot even talk to the animals, and one estimate suggests that humans can perceive only about one trillionth of the environmental signals that are present in the world. So, to use the term of the sociologist Nicholas Luhmann, human societies pick up signals (mostly using instruments they have invented, including language) from their environment and *resonate* according to the strength and amplitude of those signals.[15] Looking back at human history we can try to pick out some resonance patterns that seem to be common to

[15] N. Luhmann, *Ecological Communication* (Polity Press, Cambridge, 1989).

more than one age or to show steady gain; if, that is, we can pick them out above the year-upon-year events that constitute apparent noise in the system.[16]

One major class of constructions that we can pick out through time is that of *environmental determinism*. This concept dates from at least Classical times and revolves around the notion that the way of life of a society and its historical evolution is determined by the physical conditions of its environment. Climate is often a leading factor, but soils and landforms are also implicated in some versions. Thus the tropics are thought to breed people who are basically lazy, and southern parts of European countries to be less productive and go-ahead than their northern counterparts. Many other stereotypes have been formed and some have grains of truth in them, but all are confining both to the people thus depicted ('we can't help it') and to the depictors ('they'll never get any better: it's in their nature'). Some of the extreme forms of this type of determinism disappeared with the nineteenth century, though one advocate assumed that it was climate that made New England the manufacturing powerhouse of the USA, with the changing seasons being a stimulus to hard work and inventiveness – not a view likely to be shared by anybody stuck in a Massachusetts building in summer without working air conditioning. However, environmentalism is more often disapproved than disproved, and it came markedly to the fore again in the environmentalist movement in the western world of the 1965–72 period, rather like the recurrent march theme in one of Charles Ives's *New England Portraits*. Here, the main change was that of scale: it was the whole human species that was to be limited in its material development by environmental constraints. The underlying tenor was strongly Malthusian: rapid population growth meant that there would be limits to resource supply, limits to the capacity of the planet to process the wastes of industrial societies, and indeed limits to the planet's capacity to absorb without unpredictable fluctuations all the various changes that were being wrought in its systems. The key to it all was, largely, population growth rates and the curtailment of these by various applications of chemistry and physics

[16] C. L. Glacken, *Traces on the Rhodian Shore* (University of California Press, Berkeley and Los Angeles, 1967).

(unless Roman Catholicism was very strong, in which case only mathematics was officially allowed). After the UN's 1974 conference on population, this was soft-pedalled since (a) some rates of growth were indeed decreasing and (b) it was not thought reasonable to blame the developing countries for a world problem not of their making. Some of those countries in turn were inclined to view attempts at population limitation as yet another manifestation of imperialism. This viewpoint is currently rather *sotto voce*, substituting 'production' for population, but this might be taken as a variation on a theme rather than a turning down of the volume. The sound was turned up again a little after the 1992 Rio Conference, mostly by the UK delegation.

Another theme which we can pick up through much of the history of the West at least is that of the creation of *a humanized world within the world of nature*. Such a world has considerable reference to the natural environment, but can at times approach its virtual elimination, as we have seen in various examples through this book. A motif which appears time and time again in history is the notion of the eventual perfectibility of humanity and, in some versions, the concomitant perfection of nature, since it could not be flawless if humans were themselves less than impeccable. The rationales for these improvements were various, but tended to come to the same thing in the field: the transformation of wilderness and pristine ecosystems into cultural systems. Generally, however, the valuation of the wild was for one reason or another lower than that of the tame, and so the job of men was to listen to Isaiah:

> Every valley shall be exalted, and every mountain and hill shall be brought low: and the crooked shall be made straight, and the rough places plain

rather than the *Benedicite*:

> O all ye Green Things upon the Earth, bless ye the Lord, praise him and magnify him for ever.

In cultures which magnified the value of work, whether as beneficial to the people for whom it was done or as saving the souls of those who did it, nature tended to be viewed as a set of opportunities for individuals to change themselves, rather than allowing an innate value to nature itself: it was a mirror, so to speak, which might be polished up to show off the viewers in the brightest light.

Yet another sub-group of the seekers after perfectibility saw the world of nature as incomplete: it was in places manifestly unsuitable for human occupation, with very wet or very dry places and ugly mountains (Adam of Usk had to traverse the St Gothard Pass in 1401 and had himself blindfolded and carried across – analogous to those who on a rough day in the Channel immure themselves in the ferry's bar). The coming of science in the sixteenth century added to the strength of the movement for change, nowhere more so than in the work of one of its chief advocates, Francis Bacon (1561–1626), who insinuated the new thinking into the *Zeitgeist* by suggesting that this was the route by which the Fall might be reversed and Eden regained. In order to achieve this, *Nam et ipsa scientia potestas est*, knowledge itself is power. No wonder that his contemporary John Donne thought that the cosmos was lost:

> 'Tis all in pieces, all coherence gone
> All just supply and all relation.

Such strands unify into a broader theme when they can form the basis of a coherent ideology, as happened with the combination of science, evolutionary theory and the glorification of work in the historical materialism of the nineteenth century of which the major catalyst was Karl Marx (1818–1883). Although the valuation of nature in Marx's writings is not always consistent, it is broadly seen as a set of resources for the betterment of humankind and its transformation by the workers is what gives value to it. Thus a labour theory of value depends upon the transformation of nature and not its retention in its natural state. Marx himself declared that he was not a Marxist, but even so he would probably have approved of the Chinese in the 1960s declaring that with the sayings of Chairman Mao they could conquer nature. Modern Marxians, though, talk in terms of such philosophical idealism when they refer to the production of nature and the production of space, where both are social forms of knowledge rather than having an absolute existence. In what seems an extreme form of these trends, American commentators have remarked upon the duality of their nation's attitudes, at once transforming in the cause of material gain *and* being in the vanguard of conservation of National Parks and wilderness areas. This can be traced back to a Puritan heritage in which destruction is a necessary prelude to the seeking of a paradise: that which cannot be perfect (i.e. the individual human, projected onto the land) is destroyed rather than reformed, until virtually the last moment.

In the case of the environment, the combination of the mass nature of industrialism as adumbrated in the early nineteenth century in western Europe (and especially in the UK), together with the changes in the surroundings of the sensitive artists of the day, provoked the Romantic movement in which the individual was championed against the mass and nature against the mechanical. William Wordsworth is most often quoted, as in *Tintern Abbey* (1798), where merely to look on nature is provocative of sentiment:

> And I have felt
> A presence that disturbs me with the joy
> Of elevated thoughts . . .
>
> Whose dwelling is the light of setting suns
> And the round ocean and the living air,
> And the blue sky, and in the mind of man.

Romanticism is by no means dead: the word 'nature' still functions as a norm for some people (as in 'it's only natural'), even though they do not perceive the human-created among the natural. The enshrining of many moorland areas of England and Wales in National Parks in the early 1950s was done under the aegis of an Act of Parliament that spoke of their natural beauty, and Areas of Outstanding Natural Beauty form the second rank of protected areas – but probably none of them has a square metre of natural ecosystem.

The notion of perfectibility is not totally moribund either. Subsequent to an Act of the US Congress of 1969, most countries in the middle and upper income groups now insist on the compilation of Environmental Impact Statements for major projects. These science-based projects start with a baseline inventory of the current ecological systems (whatever their state) and then predict what would happen to them if the proposal were followed or if an alternative were to be pursued instead. The underlying current of thought is clear: it is towards a greater perfectibility of decision-making in the use of the environment as a resource.

In both determinism and perfectibility there exists one common theme: the unhesitating use of 'human' and 'environment' as separate entities, with 'man' being generally used for the first in sources unaffected by feminism. The separation of humanity from the rest of nature is so ingrained in our thinking that we rarely consider either the history or the inevitability of the conception. As with so many intellectual threads in this field, its antiquity is considerable. There seems little doubt that a good start was made by the idea of man being made in the image of God, though the Greeks were often ready to admit that the great powers of this special creature might be ambiguously used:

> Wonders are many and none more wonderful than man . . .
> Subtle beyond hope is his power of skilled invention, and with it he comes now to evil, now to good

wrote Sophocles in *Antigone*. The Renaissance built up this image of mankind as separate and indeed superior by making humans the measure of all things as part of their reversion to Classical ideals, but were overtaken by the development of scientific thinking, whose strongest statement in this area is encapsulated by Descartes' early seventeenth-century formulation, *Cogito ergo sum* – 'I think, therefore I am'. This not only separated mind from body in humans but the thinking animal (as defined by one of them) from all else. Once this separation is established as having been sanctioned then the way is clear for a dominance hierarchy of the kind 'man →→ woman →→ animals →→ plants →→ non-living materials'. Ameliorations such as those of Jeremy Bentham (1748–1832), who said of animals that the question was not 'can they think but can they suffer', seemed at the time to be only minor, though they came at the end of a period of an increasing sensibility towards the non-human.[17] But as late as the mid-nineteenth century, a pope (Pius IX) forbade the establishment of a Vatican branch of the Society for the Protection of Animals since animals had no souls and therefore were not fit subjects for moral consideration. (And indeed in 1973 the Archbishop of Los Angeles warned against 'dogmatic ecology': better, he said, to maintain the view of 'Nature as Enemy, the alien force, to be conquered and broken to man's will'.)

Above all, though, the separation has been most felt in classical economics and in technology. In the first, the highly rational nature of the discipline has a direct link with the thinking process and it is possible to consider the econosphere as if the ecosphere did not exist, a rift which today's economists sometimes try to bridge. In the second, the difference between the carbon-based world of organic evolution and the metals-and-plastics-based creations of humans is highly apparent. In particular the contrast between the slowly moving and unpredictable nature of evolution and the instant and highly (though never completely) forecastable is strong. Unsurprisingly, the latter is valued more highly by most if not all human societies except those that are consciously 'alternative'.

The end result, in terms of today, is the formulation of a distinct *world-view*. This term, one translation of the German *Weltanschauung*, can be used to denote all those elements of a culture (including its ecological relations) to which we subscribe without much if any questioning. The dominant macrocosmic view of today is usually called the western world-view and includes such elements as a belief in progress, the inevitability of material growth, the solution of problems by the application of science and technology, and, fundamentally, the assumption of human dominance wherever possible over a separate cosmos. It is the extension of this world-view in time and space which has been called 'development'. At an operational level, it seems to consist of the further consolidation of what Schnaiberg[18] calls the

[17] K. Thomas, *Man and the Natural World. Changing attitudes in England 1500–1800* (Allen Lane, London, 1983).

[18] A. Schnaiberg, *The Environment. From surplus to scarcity* (Oxford University Press, New York, 1980).

treadmill of production expansion. Capital-intensive and energy-intensive mass production is both profitable and predictable. The whole process is abetted by labour unions and the state, for whom awkward questions can always be deferred by making next year's cake a little larger. Our particular concern can focus upon the competitive nature of capitalism in speeding up withdrawals from and additions to the environment.

Of course, behind economics there lies political economy, and beyond that there are fundamental questions of values. We need to have some idea of how the environment can be and has been valued by humans and whether this leads to particular political and economic forms. There are perhaps three distinct theories about the value of nature:

1 *Anthropocentrism.* This theory lies immediately behind the capitalism described above, and many other materialist value-systems (such as that of Marx) as well. In non-human nature, all value is an instrumental value which depends upon its contribution to human value, preferably as measured in monetary terms. So nature becomes either resources or non-resources without value.

2 *Inherentism.* This theory recognizes that the very concept of value is human and so any attribution of it to nature is dependent upon human consciousness and the constructions which that makes. Nevertheless, our minds can attune to the fact that some of the values in nature do not derive from humans but from nature-in-itself.

3 *Intrinsicalism.* The notions of inherentism are taken further and we are forced to admit that some value in nature is independent of human values and of human consciousness itself. The environment has an inherent value which has nothing to do with anything human: it is to some extent a parallel world from which we ought perhaps to become decoupled. Knowledge of its values must be independent of all experience and acknowledged to be intuitively 'true'.

In today's terms, the first set of values is the ruling paradigm; the last is confined to devotees of the Deep Ecology movement; most environmentalists would sign up for the middle way but are uncertain how it can be institutionalized without tumbling over into one or other of the more extreme positions.

Together again

If we accept, none the less, that humans and nature are, as Darwin put it, 'all netted together', then we ought to consider other models of the relationship. In particular we need to think about whether history is important or whether change has been so rapid in the last few decades that all knowledge of all pre-existing conditions is obsolete.

Though we associate the natural sciences with analysis and the breaking down of phenomena into their constituents (*reductionism*) for study, there is also a tradition of the study of wholes. This *holism* is often associated with ecology and in that branch of the natural sciences the separations between living and non-living, plant and animal, human and non-human are regarded as unhelpful. In modern ecology they are replaced by the notion of a dynamic system in which they all exchange matter and energy, and in effect are all components of a web. The investigation can then emphasize links rather than differences; some workers have used energy flows for this purpose. In this manner, the flux of solar energy and the subsidies of refined oil (in say agriculture) can be considered in the same breath and the same quantitative units. Nevertheless it is true to say that some ecological work has a distinct air of viewing human society as an intruder and despoiler of the ecologist's central object of value, a natural or natural-looking ecosystem. The field of 'stress ecology' is one such set of implicit views.

This kind of ecology is mostly about relations in space, although an historical dimension may be included in studies. Following from current ecological studies, there is the kind of evolutionary narrative which asserts that life has produced many of the physico-chemical conditions of the planet, such as the gaseous composition of the atmosphere and the salinity level of the oceans. In some versions, climate is to some extent regulated by life. In this scenario, if humans contravene the laws laid down by the very existence of life then they will probably be expelled from the stage. This world-view, at odds with the western

Weltanschauung described above, is usually termed the Gaia hypothesis, named after the Greek goddess of the earth by its formulator, J. E. Lovelock. Integration into a more humanistic framework is sometimes seen in the recurrent myth of a Golden Age. This is present in most branches of human culture, but nowhere more so than in the concept (present at least since the time of Genesis and of Classical Greece) that there was a time in the past when all was lovely, in the Garden among other places. Thus in that time, humans lived in beautiful places, at peace with each other and with nature, and there were none of the problems humankind has had to face since. It is a myth reformulated at intervals throughout human history: in Classical times, the Epicureans thought that the earth had as many people and crops as it could support, and in any case was basically senescent: the good days were over; In the 1970s a National Park planning committee in England was likely to have on it a group of people whose image of the region was that of the 1920s, with colourful but deferential farmers, steam railways and only a few non-natives (themselves, as it happened) with a special sensitivity for the place. (This was certainly true in the Yorkshire Dales and the golden vein was richly tapped in the James Herriot books and TV series.) Myth can indeed integrate humans and nature, but not always in a way which appeals to those charged with planning for the future.

Both Gaia and Golden Age seem to be examples of what science is often looking for: the presence of laws. Equally, those who study the past are always alert to the possibility that there has always existed in human behaviour the same kinds of regularities which would qualify for the label. Assertions about the corrupting tendencies of power or the universality of Murphy's Law perhaps fall into this category. On the other hand, there is the possibility that in human affairs there is no such thing as the inevitabilities (and certainly not the predictabilities) implied by the concept of laws in the scientific sense. All is contingent in the manner suggested by Gould:[19] what happens next is simply a consequence of what happened immediately beforehand and what the local cir-

[19] S. J. Gould, *Wonderful Life. The Burgess Shale and the nature of history* (Hutchinson, London, 1989).

cumstances are. Gould argues winningly, for example, that the evolution of any species (including our own) is more or less an accident of a particular time and place. Adoption of such ideas has a dethroning impact upon human pretensions and can reassert the view that nature has an intrinsic value of its own. The non-human world should then have a worth which transcends that put upon it by humans, since at heart their price will reflect the utility of nature to humans and not its worth as something-in-itself. We can see a remnant of that outlook in the environmentally tender statements by North American aboriginal chiefs reproduced on T-shirts and posters.

Possibly the past is not such a different country. We cannot know it directly, but we cannot know the present directly either. As a species we cannot perceive more than a tiny fraction of what is around us, even with complicated instruments (which in any case have been designed and built by us so they do not extend objectivity very significantly), and so much is comprehended if at all in the form of metaphor. To get closer to the true nature of the past in its significance for human-environmental relations is therefore a mammoth task and one in which the local and the contingent will always be as important as seeking for the Grail of law-like generalizations.

It seems as if both law and contingency are involved and important. At one temporal scale of evolution the former seems to govern. This is the cosmic scale in which the laws of thermodynamics have no loopholes either for humans or the rest of energy and matter. The arrow of time does not appear to be reversible and entropy is continuously being created from more concentrated sources of energy. (We *could* argue from there that if we are all going to hell in a bucket one day then what harm is there in speeding up the process a bit by using the fossil fuels and aerosol sprays as if there was no tomorrow.) Life and human societies are temporary aberrations in the creation of entropy. They are structures which take in highly concentrated energy and use this to create complexity before passing it on largely as heat, unable to do any further work, merely a contribution to the ultimate random distribution of everything in the universe.

One further lesson seems to be that the universe is far from equilibrium, and that all energy and matter cascades towards a high-entropy state in ways which we cannot predict and possibly will never be able to foretell accurately. In the science of ecology, as we have already seen, the certainties of succession-climax theory have already given way to a more probabilistic view in which outcomes are much less predictable in any detail, rather like weather forecasting beyond the next day or so. At the global scale of human-environment interactions, it seems unlikely that there is any trend which is sufficiently long to make any reliable prediction for anything other than the most simple world-wide processes.[20]

The outcome of this type of thinking (currently labelled as chaos theory) is that even in the presence of laws such as those of thermodynamics and gravity there are still open futures because of the presence of complexity and the ways in which a system can be self-organizing. It is here, presumably, that the smaller-scale processes of contingency become important. They do not contravene the overarching conditions of gravity and thermodynamics, but they do introduce a great deal of indeterminacy at the local scale and life is especially important in this respect. The kind of systems which have developed (including those in which humans and their cultures are involved) are clearly dependent for their coherence upon the controls which keep them from behaviour which is totally lacking in any form of cohesion and which therefore is too terrible for us to contemplate: imagine an environment in which the climate was totally unpredictable (within any known limits) from year to year. These controls (known as feedback loops) present themselves in the form of environmental processes and our impacts upon them affect their evolution and their history, which must be known and given meaning.

The totality of these rather large-scale ideas is, however, quite simple: we are creatures of evolution on cosmic, ecological and cultural scales. We are also creators on the last two of these. But in none is there a break-point which would enable us to ignore what had gone before. History is like a narrative tapestry: if we cut it up and store some of it in a chest, we shall not understand the message of what is left for us to see hanging on the wall.

[20] R. A. Treumann, 'Global problems, globalization, and predictability', *World Futures*, 31 (1991), pp. 47–53; D. Worster, 'Ecology of order and chaos', *Environmental History Review*, 14 (1992), pp. 1–18.

Guide to Further Reading

Environmental histories which concentrate on scientific evidence and are therefore strongest on the earlier part of the Holocene include N. Roberts, *The Holocene. An environmental history* (1991), A. Mannion, *Global Environmental Change. A natural and cultural history* (1991) and M. Bell and M. J. C. Walker, *Late Quaternary Environmental Change. Physical and human perspectives* (1992). The grand-daddy of a more humanistic approach was W. L. Thomas (ed.), *Man's Role in Changing the Face of the Earth* (1956); it is succeeded by B. L. Turner et al. (eds), *The Earth as Transformed by Human Action. Global and regional changes over the past 300 years* (1990), though that only covers the period specified. My own *Changing the Face of the Earth. Environment, history, culture* (1989) tries to balance both scientific and humanistic approaches. The newer discipline of environmental history *per se* is addressed in books such as P. Brimblecombe and C. Pfister (eds), *The Silent Countdown. Essays in European environmental history* (1990), and an excellent 'popular' treatment is in C. Ponting, *A Green History of the World* (1991). Pfister's treatment of the environmental history of the Büren region of Switzerland on an ecological-energetics basis in *The Silent Countdown* is a very interesting and quantifiable approach.

The leading periodical with a humanist approach is the journal of the American Society for Environmental History, *Environmental History Review*. There is as yet no European equivalent. Factual material is therefore spread among a number of sources in economic and social history and in historical geography; for 'prehistoric' periods, journals of Quaternary ecology often contain useful material.

There is a burgeoning literature in the social science of the environment, dominated perhaps by political scientists, but with most others joining in. They have been joined by a few philosophers: J. Passmore's *Man's Responsibility for Nature* (1982) has become the book that cannot be ignored, though not always agreed with. The philosophy of environment in general has become the special province of the journal *Environmental Ethics*, spilling beyond the confines of that title. In terms of metatheoretical explanations of environmental change at human hands, I have found the economic model of A. Schnaiberg, *The Environment. From surplus to scarcity* (1980) the most convincing, while feeling that single-factor explanations are almost always inapplicable in this field.

In general, though, the literature is scattered and courses in higher education have usually to collect together a wide and eclectic diversity of material, rather than mining a single and well-defined seam. A nuisance for the hard-pressed teachers but probably a good thing for the students.

Bibliography

Albion, R. G., *Forests and Sea Power. The timber problems of the Royal Navy 1652–1862*, Harvard University Press, Cambridge, MA, 1926.

Astill, G. and A. Grant (eds), *The Countryside of Medieval England*, Blackwell, Oxford, 1988.

Auclair, A. N., 'Ecological factors in the development of intensive-management ecosystems in the mid-western United States', *Ecology*, 57 (1976), 431–44.

Bamford, P. W., *Forests and French Sea Power 1660–1789*, University of Toronto Press, Toronto, 1956.

Bell, M. and M. J. C. Walker, *Late Quaternary Environmental Change. Physical and human perspectives*, Longmans, London, 1992.

Birch, T. H., 'The incarceration of wilderness: wilderness areas as prisons', *Environmental Ethics*, 12 (1990), 3–26.

Bratton, S. P., 'The original desert solitaire: early Christian monasticism and wilderness', *Environmental Ethics*, 10 (1988), 31–53.

Brimblecombe, P., *The Big Smoke*, Methuen, London, 1987.

—— and C. Pfister (eds), *The Silent Countdown. Essays in European environmental history*, Springer, Berlin and Heidelberg, 1990.

Clark, J. D. and J. W. K. Harris, 'Fire and its roles in early hominid lifeways', *African Archaeological Review*, 3 (1985), 3–27.

Clarke, K., *Landscape into Art*, John Murray, London, 1949.

Colinvaux, P., 'Amazon diversity in light of the palaeoecological record', *Quaternary Science Reviews*, 6 (1987), 93–114.

Crosby, A. W., *Ecological Imperialism. The biological expansion of Europe 900–1900*, Cambridge University Press, Cambridge, 1986.

Cushing, D. H., *The Provident Sea*, Cambridge University Press, Cambridge, 1988.

Delano Smith, C., *Western Mediterranean Europe. A historical geography of Italy, Spain and southern France since the Neolithic*, Academic Press, London, 1979.

Diamond, J. R., 'Human use of world resources', *Nature, Lond.*, 328 (1987), 479–80.

Flenley, J. R., *The Equatorial Rain Forest: A geological history*, Butterworth, London and Boston, 1979.

—— et al., 'The late Quaternary vegetetational and climatic history of Easter lsland', *Journal of Quaternary Science*, 6 (1991), 85–115.

French, R. A., 'Russians and the forest', in J. H. Bater and R. A. French (eds), *Studies in Russian Historical Geography*, Academic Press London, vol. 1, 1983, pp. 23–44.

Glacken, C. L., *Traces on the Rhodian Shore*, University of California Press, Berkeley and Los Angeles, 1967.

Gould, S. J., *Wonderful Life. The Burgess Shale and the nature of history*, Hutchinson, London, 1989.

Green, D. G., 'Fire and stability in the postglacial forests of southwest Nova Scotia', *Journal of Biogeography*, 9 (1982), 29–40.

Heathcote, R. L., *The Arid Lands: Their use and abuse*, Longman, London and New York, 1983.

Holzner, W. et al. (eds), *Man's lmpact on Vegetation*, Junk BV, The Hague, Boston and London, 1983.

Houston, J. M., *The Western Mediterranean World. An introduction to its regional landscapes*, Longman, London, 1964.

Irland, L. C., *Wilderness Economics and Policy*, Lexington Books, Lexington, MA and Toronto, 1979.

James, S. R., 'Hominid use of fire in the Lower and Middle Pleistocene', *Current Anthropology*, 30 (1989), 1–26.

Lesslie, R. G., B. G. Mackey and K. M. Preece, 'A computer-based method of wilderness evaluation', *Environmental Conservation*, 15 (1988), 225–32.

Lewis, H. T., 'Maskuta: the ecology of Indian fire in Northern Alberta', *Western Canadian Journal of Anthropology*, 1 (1977), 15–52.

—— and T. A. Ferguson, 'Yards, corridors and mosaics: how to burn a Boreal forest, *Human Ecology*, 16 (1989), 57–78.

Luhmann, N., *Ecological Communication*, Polity Press, Cambridge, 1989.

McCloskey, M. and H. Spalding, 'A reconnaissance-level inventory of the amount of wilderness remaining in the world', *Ambio*, 18 (1989), 221–7.

Mannion, A., *Global Environmental Change. A natural and cultural history*, Longman, London, 1991.

Merchant, C., *Ecological Revolutions. Nature, gender and science in*

New England, University of North Carolina Press, Chapel Hill and London, 1989.

Nash, R., *Wilderness and the American Mind*, Yale University Press, New Haven and London, 3rd edn, 1982.

Nicholas, G. P. (ed.), *Holocene Human Ecology in Northeastern North America*, Plenum Press, New York and London, 1988.

Oelschlager, M., *The Idea of Wilderness. From prehistory to the age of ecology*, Yale University Press, New Haven and London, 1991.

Ornstein, R. and P. R. Ehrlich, *New World New Mind. Changing the way we think to save our future*, Methuen, London, 1989.

Pacey, A., *Technology in World Civilization. A thousand-year history*, MIT Press, Cambridge, MA, 1990.

Passmore, J., *Man's Responsibility for Nature*, Duckworth, London, 2nd edn, 1982.

Patterson, W. A. and K. E. Sassaman, 'Indian fires in the prehistory of New England', in G. P. Nicholas (ed.), *Holocene Human Ecology in Northeastern North America*, Plenum Press, New York, 1988.

Pennycuick, C. J., *Newton Rules Biology. A physical approach to biological problems*, Oxford University Press, Oxford, 1992.

Pfister, C., 'The early loss of ecological stability in an agrarian region', in P. Brimblecombe and C. Pfister (eds), *The Silent Countdown. Essays in European environmental history*, Springer, Berlin and Heidelberg, 1990, pp. 37–55.

Ponting, C., *A Green History of the World*, Sinclair-Stevenson, London, 1991.

Pyne, S. J., *Fire in America. A cultural history of wildland and rural fire*, Princeton University Press, Princeton, NJ, 1982.

Rackham, O., *The History of the Countryside*, J. M. Dent, London and Melbourne, 1976.

—— *Ancient Woodland. Its history, vegetation and uses in England*, Edward Arnold, London, 1980.

Rasmussen, P., 'Leaf-foddering of livestock in the Neolithic: archaeobotanical evidence from Weier, Switzerland', *Journal of Danish Archaeology*, 8 (1989), 51–71.

Roberts, N., *The Holocene. An environmental history*, Blackwell, Oxford, 1991.

Russell, E. W. B., 'Indian-set fires in the forests of the Northeastern United States', *Ecology*, 64 (1983), 78–88.

Schnaiberg, A., *The Environment. From surplus to scarcity*, Oxford University Press, New York, 1980.

Schüle, W., 'Anthropogenic trigger effects on Pleistocene climate?', *Global Ecology and Biogeography Letters*, 2 (1992), 33–6.

Shepard, P., *Man in the Landscape. A historic view of the esthetics of nature*, Knopf, New York, 1976.

Simmons, I. G., *Changing the Face of the Earth. Environment, history, culture*, Blackwell, Oxford, 1989.

Stankey, G. H., 'Beyond the campfire's light: historical roots of the wilderness concept', *Natural Resources Journal*, 29 (1989), 9–24.

Swain, A. M., 'A history of fire and vegetation in Northeastern Minnesota as recorded in lake sediments', *Quaternary Research*, 3 (1973), 383–96.

Thirgood, J. V., *Man and the Mediterranean Forest. A history of resource depletion*, Academic Press, London, 1981.

Thomas, K., *Man and the Natural World. Changing attitudes in England 1500–1800*, Allen Lane, London, 1983.

Thomas, W. L. (ed.), *Man's Role in Changing the Face of the Earth*, Chicago University Press, Chicago, 1956.

Totman, C., *The Green Archipelago. Forestry in preindustrial Japan*, University of California Press, Berkeley, Los Angeles and London, 1989.

Thompson, W. I., *Imaginary Landscape. Making worlds of myth and science*, St Martin's Press, New York, 1989.

Treumann, R. A., 'Global problems, globalization, and predictability', *World Futures*, 31 (1991), 47–53.

Tucker, R. P. and J. F. Richards (eds), *Global Deforestation and the Nineteenth-Century World Economy*, Duke University Policy Studies, Durham, NC, 1983.

Turner, B. L. et al. (eds), *The Earth as Transformed by Human Action. Global and regional changes over the past 300 years*, Cambridge University Press, Cambridge, 1990.

Watson, A. M., *Agricultural Innovation in the Early Islamic World. The diffusion of crops and farming techniques 700–1100*, Cambridge University Press, Cambridge, 1983.

Wein, R. W. and D. A. Maclean (eds), *The Role of Fire in Northern Polar Ecosystems*, Wiley for SCOPE, vol. 18, Chichester, 1983.

Williams, M., *Americans and Their Forests. A historical geography*, Cambridge University Press, Cambridge, 1989.

—— (ed.), *Wetlands: A threatened landscape*. IBG Special Publications 25, Blackwell, Oxford, 1990.

Winchester, A. J. L., *Landscape and Society in Medieval Cumbria*, John Donald, Edinburgh, 1987.

Woodell, S. R. J. (ed.), *The English Landscape. Past, present and future*, Oxford University Press, Oxford, 1985.

Worster, D., 'Ecology of order and chaos', *Environmental History Review*, 14 (1992), 1–18.

Index

Note: Page references in italics are to illustrations